Original edition 1951 Reprinted with changes 1955

PREFACE

This book is written for men of the Navy and of the Naval Reserve who are studying for advancement to the rates of Construction Electrician's Mate first or chief. Combined with the necessary practical experience and a thorough review of the applicable Navy Training Courses, the information in this training course will help prepare the reader for advancement in rating examinations.

Qualifications for these Construction Electrician's Mate rates are Usted in appendix II at the back of this book. Because examinations for promotion are based exclusively on these qualifications, it is suggested that you refer to them frequently for guidance.

The first two chapters discuss principles of leadership and the type of administrative duties for which a CE first or chief is responsible. Chapter 3 tells how to read and interpret electrical drawings. The fourth chapter describes the most common types of electrical repairs. Chapter 5 contains a discussion of advanced base communications equipment, including telephone systems, inter-office communication equipment, and public address systems. Telephone cable splicing is covered in chapter 6. Chapter 7 discusses transformer installation and repair. Electric power stations and equipment foimd at advanced bases are described in chapters 8 and 9. Chapter 10 discusses lead-sheathed power cable.

As one of the Navy Training Courses, this book was prepared by the Navy Training Publications Center for the Bureau of Naval Personnel, with technical assistance from the Bureau of Yards and Docks.

Hi

STUDY GUIDE

The table below indicates which chapters of this book apply to you rating. To use the table, follow these rules:

1. Select the column which applies to your rating. If your are in th Regular Navy, you will use the column headed "CE" which is th general service rating. If your are a member of the Naval Reserve you will use the column headed by your particular emergency serv ice rating— CEG, CEP, or CEL.

2. Observe which chapters have been marked in your rating coluini with the number of the rate to which you are seeking advancement

3. Study those particular chapters. They include information whicl will assist you in meeting the qualifications for your rating. (Se app>endix II of this book for a complete list of qualifications for ad van cement in rating.) In order to gain a well-rounded view of thi duties of the general service rating, it is recommended that you rea< the other chapters of this book even though they do not pertaii directly to your rating.

4. Here is an example: If you are a member of the Naval Reserve studying for advancement in ratmg to Construction Electrician^ Mate G (General Electricians), you will select the column headed CEG. Following this column down you will observe that you mufi study chapters 1, 2, 3, 4, 7, 8, 9, and 10.

I

CONTENTS

Page

CHAPTER 1

V

READING LIST

NAVY TRAINING COURSES

Conatruction Electrician's Mate S <k S, XavPers 10636-A
Electricity, NavPers 10622-B
Electrician's Mate 1 & C, NavPere 10550-A
Basic Hand Tool SkiUs, NavPers 10085
Blueprint Reading, NavPere 10077

OTHER PUBLICATIONS

Jan Standard 15

USAFI TEXTS

United States Armed Forces Institute (USAFI) courses for additional reading and study are available through your Information and Education OflScer.* A partial list of those courses applicable to your rate follows:

Number Title S«lf Teaching
MA 520 Electricity and Magnetism
MB 835 Machine Shop 1
MB 785 Electrical Measuring Instruments

MA 890 Principles & Practices of Radio Servicing
MA 885 Fundamentals of Radio
Corrttpondcncc
CA 885 Fundamentals of Radio CB 835 Machine Shop I

♦"Membere of the United States Armed Forces Reserve Components, when on active duty, are eligible to enroll for USAFI courses, services, and materials, if the orders calling them to active duty specify a period of 120 days or more, or if they have been on active duty for a period of 120 days or more regardless of the time specified in their active duty orders."

CONSTRUCTION ELEaRICIAN'S MATE 1 & C

CHAPTER 1

SEHING YOUR SIGHTS NEW RESPONSIBILITY

Those stripes on your arm show how far you have come up the ladder since you were a striker. They indicate that 3"ou have learned your job well and mark you as an expert in the field of electrical equipment. You are now ready to set your sights on a higher goal—First or Chief.

Previously, in studymg for advancement, you have concentrated upon increasing your technical skills and developing the knowledge essential to those who work with electrical equipment. Throughout your naval career, you must continue to study and to develop your knowledge and skills. As a Fii"st or Chief, you will liave the additional responsibility of supervisor\'7d^ duties. A successful supervisor must be a leader, organizer, disciplinariaji, and teacher.

You learned long ago that you need actual experience to master a skill or a technique. You will find, likewise, that you need experience to become a good supervisor. But, just as you learned some of the fundamentals of electricity from textbooks, so you can learn some of the principles of

supervision from the experience of others. The following pages contain a discussion of some of the basic principles ofi good supervision.

ROUTINE AND DISCIPLINE

The function of a leader is to organize the activities of the group toward the accomplishment of a given task. In any good organization, the duties and responsibihties of each member are spelled out in detail. Each man should know exactly what is expected of him. However, you knov from past experience that listing duties does not necessarily get them accomplished. The established routine of work activity must be reinforced by discipline. For example, i powerplant generates electric power because its many pieces of equipment are regulated, or disciplined to work in unisos toward a specific objective. Because you are dealing with human beings and not with inanimate objects, your problems of supervision and discipline are more complex. You must supervise in a calm and understanding manner to get prompt and willing compUance with orders. You will get more respect and better results by being reasonable and fair than in any other way. Of course, fairness must not be confused with laxity or undue leniency.

In preparing man-hour schedules, remember that yo«i men will need occasional breaks and periods of relaxation. If you do not take this factor of fatigue into consideration, your methods will tend to become tyrannical. On the other hand, if you grant an excessive number of breaks, you may lose control over the group. In each individual situation, strive for a happy medium. Be flexible enough to make allowances for changing conditions. Make the standards of conduct imiform for everyone working imder your supervision. In giving praise or censure, never show favoritism; be impartial, consistent, and humane.

PRAISE AND CENSURE

There is a common misconception that discipline is merely a matter of administering censure or pimishment. Disci^

pline also has positive aspects that are frequently overlooked. When inspecting the quality of work done by one of your men, resist the tendency to comment only on the mistakes that have been made; occasionally, you should conmient favorably on the good parts of the job. Here too, you must use good judgment. No one is entitled to praise for simply doing his job. If you pass out commendations indiscriminately, they soon lose their meaning. A good rule to remember is that a kind word boosts morale; flattery tends to destroy it.

When praise is handed out from your superiors, be sxu-e to pass it on down to your men who helped earn the commendation. Recognition given to your crew for a job well done is music to their ears.

Another rule to remember is that it is well to praise in public, but to censure in private. Be certain that a man understands why he is being censured. Be especially careful to avoid sarcasm.

KNOW YOUR MEN

It is important to know and understand your men. This understanding is essential if you are to treat them as individuals and not merely as cogs in a machine. Learn all you can about each member of your crew—his name, ambitions, problems, capabilities, and limitations. Show the men that you are sincerely interested in their welfare. They want a supervisor who is interested in them and to whom they can look for approval. If you know and understand them, have respect for their dignity as individuals, and show a sincere interest in their welfare, they will respond with renewed energy, pride, and confidence in your leadership. They wiU work because they want to, and not because they are compelled to. Such a working relationship is sometimes called voluntary discipline. It is based on knowledge, reason, sense of duty, and loyalty; it is closely related to esprit de corps.

At the same time you are learning to know your men, give them an opportunity to get to know you. When they have work to do, you should put in an appearance to let them see

that you are interested in their work. A word of caution: Don't make the mistake of constantly standing over your men and breathing down their necks. Such conduct tends to make them self-conscious and to lessen confidence in themselves.

Your men will depend upon you to fulfill many of their psychological needs. By interviewing them and by chatting informally, you can show each man that you are interested in him as an individual. Be very careful, however, not to pry into private affairs. When a man comcs to you with a personal problem, listen sympathetically, but be careful about giving advice. It is not your job to become a chaplain in dungarees.

PASSING THE WORD

Every person in the Navy must obey the order of his lawful superior or become liable for punishment. It is easy for a supervisor to operate entirely on this basis; that is, the supervisor can give orders, make no explanation as to the purpose of the orders, and require blind obedience. To a certain extent, this is necessary. No supervisor has time to explain the necessity for all of his orders. It should never be forgotten, though, that men do better work when they understand the purpose of an order. When they feel the job they are doing is important, they ignore discomfort and danger and put forth extra effort. How do you make a group of men feel that their work is important? The answer is by showing them that their job is an essential part of something big. For this reason, you should keep your men well informed about the general plans of the activity of which you are a part. When orders or instructions are issued to you by your superiors, pass the word on to your men. Show them the reason for their duties, and make clear to them how those duties fit into the overall picture.

Passing the word is not a one-way proposition; it is just as important that the word be passed upward as downward. Be on the alert to get your men's reaction to any situation

that arises. Encourage them to confide in you. Small, petty gripes have a way of gnawing at men in a group and playing havoc with their morale. Unless you are aware of the gripe, you are in no position to take remedial action. Very often, if you know the facts, you will be able to correct the situation. Naturally, you can't work miracles. Even so, your men will feel better if they realize that you know their feelings. Be careful not to make promises that you may not be able to keep; a broken promise causes loss of respect.

GIVING AN ORDER

As First or Chief, you will be responsible not only for your own work but also for the work of others. Your main job will be to get your men to do their jobs with maximum effectiveness. To accomplish this, you must issue orders.

For a long time you were on the receiving end of orders; you know from experience that some orders are easier than others to take and to obey. Think about it, and you will recall the three main. characteristics of an easy-to-follow order. It is clear, concise, and complete, and is issued in a pleasant manner. Don't forget that the why of an order often pays off in good will and cooperation.

In giving orders, treat your men as you would like to be treated. The men under your supervision will experience reactions similar to yours when you were on the receiving end. The following are specific suggestions for the application of sound leadership principles when giving orders:

1. Think your order through before you give it. Make a decision and stick to it. Adding afterthoughts can result in hopeless confusion.

2. Explain what you want done so that there can be no misunderstanding. Recheck to confirm that your order was understood correctly.

3. Show that you have confidence in the ability of your men to get the job done. It is not necessary to spell out in detail how to do the job unless the individual asks for such assistance. Encourage your men to use their initiative.

4. Be available to encourage and to coach your men if they encounter difficulties.

5. Avoid talking down to the men.

6. Avoid an overbearing attitude in giving instructions. Remember that the men are servii^ the Navy, not you personally.

7. In giving an order, try to get across the feeling of "let's go" instead of "get going,"

Giving an order is only part of the story. Be sure to provide a subsequent followup or check to see that there is prompt compliance. Sound leadership ability of the sup.ervisor combined with the disciplined obedience of his men produces high morale and peak efficiency.

DELEGATING AUTHORITY

In your capacity as supervisor, you may find your work load growing larger and larger. A supervisor who overburdens himself with endless detail may become as ineffectual as one who lets matters take their course. When the number of men, the time factor, or the distance, limits your control of a situation, or if you find yourself boding down in a sea of small details, the time has come to delegate some of your authority. Delegating authority to others is a privilege of supervisors, but like most privil^es, one to be used with caution. Seniority is just as important going down the line as going up; consequently, when delegating authority, you should select the man next senior below you in the chain of conmiand. If you bypass this man, you will cause resentment and defeat the purpose of your organizational structure.

Make certain that any man to whom you have delegated authority clearly understands what is to be done and the extent of his authority. Let him exercise his judgment in working out the details of how the job should be done. If he needs help, encourage him to come to you for advice. Check with him from time to time to see how the job is coming along, but don't get in his way.

You must remember that delegating authority does not relieve you of final responsibility for seeing that a task is accomplished satisfactorily; this is primarily a means of sparing you unnecessary details. It cannot be used as a method of "passing the buck."

Delegating authority serves still another useful purpose. It is a good method of allowing you to observe and develop the supervisory skill of your senior men.

HUMAN RELATIONS

Human relations is the rather fancy name given to the art of getting along with people. There is no particular set of rules that will guarantee your success in getting along with everybody. Nevertheless, there are certain character traits that you can develop to make the job easier. The most important of these are dependability, punctuality, consideration, and tact. Dependability means reUabihty or trustworthiness. Punctuality means promptness; being where you are supposed to be on time. Consideration is the thoughtful regard and respect for the needs and feelings of others. Tact means presenting the truth with consideration; it is the knack of saying the right thing at the right time, without giving offense. It is difficult to be tactful at times, especially when the situation is urgent and tempers have been roused. The technique of giving the other fellow a chance to get out of an argument gracefully is one of the keys to good human relations. There are other things you can do to help yourself get along better with people. Learn to smile in a sincere, warm, and friendly manner. Become genuinely interested in other people. Try to make the other person feel important, and do it sincerel\'7d'-.

The problem of leadership and successful human relations is not a question of knowing a long list of specific techniques; it is a question of knowing when to employ each technique. Good common sense is your best guide in this matter.

Getting along with people is a knack that can be acquired,

but, as with any other skill, it takes practice. You will find many pointers in General Training Course jor Petty Oficers, NavPers 10055. You must remember that unless you apply these pointers to everything you do and say, 7 days a week, they won't help you any more than if you had never opened the book. Nothing succeeds like success. If you make it a practice to handle effectively all the minor problems in your daily work, you will find that you will have fewer major problems. Moreover, those that do arise will be easier to solve because of your previous successful experiences in handling people.

PLANNING AND DIRECTING THE WORK

One of the most important factors in supervision is a job plan. Without a clear-cut plan of attack, you will often find that you are merely muddling through the assignment. Here are the factors to consider in setting up a job plan: (1) what the job is; (2) how you are going to accomplish it; (3) how many men are required for the job; and (4) what tools and materials are necessary. Schedule the activities so that your men work in proper sequence. This requires planning the work so that as the men complete the first part of the job, they can systematically move on to the second part. If you have two groups working on one job, you must plan the timing very carefully. This is especially true when your men are working on a job with other ratings.

In distributing work, be fair to all the men. It is a natural inclination, and a part of every person's makeup, to give the i breaks to people he likes. The important thing is to realize j that you have this inclination and to control it. Favoritism is one of the chief causes of low morale and poor production. Let ability be the measure of your men. When you have an especially good detail to fill, think carefully before making your choice. Choose the man best qualified for the job whether you like him or not. Never give your men reason to believe that the only way to get ahead is to become your pal.

Whenever possible, assign jobs that will challenge the initiative and abilities of your men. This will give you some idea of the capabilities of the members of your crew and will help you in planning a training program.

TRAINING AND INSTRUQING

As a leading petty officer, you must be on the alert to spot the training needs of your men. When new methods, equipment, or jobs are introduced, or when promotions or transfers are made, you will frequently find a specific need for training. New men may lack adequate skill for the specific jobs assigned them. Sometimes they need refresher training.

The Navy relies heavily on the higher rated petty officers for instruction and training on the job. One of your biggest problems in connection with instruction and training is keeping yourself informed on the individual training status of each man. Before beginning any training program, you should have answers to the following questions:

1. What is the job that you want to teach?
2. How much or how little training is required?
3. Who should do the training?
4. When should it be done?
5. How can it be done best?
6. Where should it be carried out?

In developing a specific plan for on-the-job training, you may find it useful to use a training timetable similar to the one in figure 1-1. The timetable should list all your men and all the jobs performed by your unit. Then place a checkmark for each man under the jobs he can do satisfactorily. Opi)osite the job that does not apply to the man in question, place a dash. For jobs that you want the man to learn, determine the date by which you feel he should be trained for the job. Opposite each name you will have a check, a dash, or a date.

In preparing a training timetable, consider the workload of your unit, the approximate future workload, the abilities and interests of your men, and their present work performance. It is well to remember that you may have personnel changes,

326605°—65 2 9

Figure 1-1.—Sample trainins fimetable.

such as promotions and transfers. Space the training so that it is neither crowded into a short period nor extended over a long period of time.

METHODS OF TEACHING

There are many methods of teacliing—lecture, demonstration, discussion, pupil-coach, and laboratory or shop work. In your training program, you will have occasion to use all of these methods. The Manual Jor Navy Instructors, Nav-Pers 16103-B, is an excellent reference on the techniques of teacliing.

Through formal training (lecture and discussion) you can teach your men the theories and principles that are essential to those who work with electrical equipment. Through demonstrations you can show them the right way of doing a job. You will find, however, that telling is not always teach-

ing, observing is not always understanding, and listening is not always learning. A trainee learns best by doing. On-the-job training is the best way to produce real skill.

The following procedure is suggested for an on-the-job method of instruction:

1. Explain the importance of the job and how it fits in with the rest of the work.

2. Show the trainee how to do the job, explaining each step as it is performed. Put special emphasis on safety.

3. Be sure the trainee understands how to do the job before he begins.

4. Check the trainee's progress carefully so that mistakes can be corrected immediately.

5. Provide opportunity for supervised practice to make sure that undesirable work habits are not developed.

6. If undesirable work habits occur, reteach that portion of the job.

When the trainee is able to do the work satisfactorily, leave him on the job to gain confidence and experience. Come back later and examine his work to be siu*e he is doing what is required. As he becomes more accomplished, you can check less frequently.

THE TEACHER IN YOU

EflFective teaching requires many of the same qualities as leadership. Your technical ability, personaUty, enthusiasm, sense of humor, sense of responsibility, ingenuity, and military bearing will all enter into the picture. The list of characteristics of a good instructor is endless, because each teaching situation, as well as each individual learner, is different. With one learner, you may have to be hardboiled; with another, you may have to be especially patient; and with a third, you may have to be tactful. In any case, you must be adaptable enough to adjust yourself and your methods to the requirements of the learner, the subject matter, and the situation. You must motivate the trainee to want to learn. Arouse his interest, desire, and willingness, to make him ready to learn. Make him see the practical use of the subject or skill he is expected to learn.

Learning experiences should be vivid and should appeal to the senses. A trainee learns through his senses, especially the sense of sight. You could talk all day about an electrical circuit in a telephone, but unless you helped the learner to visualize what you have told him, he would have only a vague notion of the subject. A simplified chart or diagram usually goes a long way toward clarifying a complex idea or principle.

The more senses a trainee uses while learning, the faster he will learn. For example, a training film with sound will produce a more vivid and lasting impression than a silent film. Similarly, a training situation appealing to the senses of touch, sight, and sound will be far more effective than one stimulating only one of the senses.

Training is performed to develop either a pattern of thinking or a pattern of acting. In either case, it is important that the trainee take an active part in the proceedings. This is best accomplished through on-the-job training.

In any teaching situation, there will be times when you will not know all the answers. At such times, the proper procedure is to admit that you do not know. If the question is important, however, you should take some action toward finding the correct answer. When you are lecturing or working with a group of men, ask the group if anyone knows. If that fails, you may assign the problem to a group member or offer to look up the answer yourself. Do not attempt to bluff your way through the situation. You may not get away with it. If you do, the traiuee may get a wrong (and possibly dangerous) impression. In the long run, bluflSng does not pay. '

SAFETY j

Heretofore, your interest in safety has been to keep yourself and others from getting an injury through one of your own acts. As the ranking petty officer, your responsibilities for safety are greatly increased. The man in charge of a work detail is responsible for the safety of his men. Never overlook the safety factor in making a decision or giving an order. Each man does his job the way you tell him to do it.

If you do not tell him how to do it, he will select his own method; he will assume that his method is acceptable unless you instruct him to the contrary. You must keep a close eye on the practices and procedures of your men.

The safety of personnel is a major concern of the Navy. The bureaus of the Navy Department from time to time have published specific safety precautions in the form of directives. These directives were supplemented with additional information, systematized, and published in a single volume. United States Navy Safety Precautions, OpNav 34P1. Chapter 18 discusses safety in working with electricity and electronic equipment and contains many valuable tips on safety procedures. Study the copy at your activity, and follow tjbe recommendations contained therein.

WORKING AS A TEAM

As a third class you learned about organization in a Construction Battalion. Good supervision depends upon the firm foundation of a well-planned organization. Organization provides the structure through which plans are formulated, lines of communication are established, authority is delegated, quality and quantity of work are achieved, materials are procured, and missions are accomplished. Teamwork, perhaps the most important factor in carrying out an operation, is greatly facilitated by good organization. Consider the teamwork necessary to build a concrete-decked mess hall.

First, the Surveyor must place the stakes for the location and elevation of the building. Next, the Driver clears the location, does the necessary excavations, and hauls in the building material. A \Iechanic lubricates and services the heavy construction equipment used in the operation. A Builder pours the concrete and performs the actual construction of the building. Before he pours the concrete, a Construction Electrician's Mate lays the conduit for the electrical wiring. A Steelworker adds his efforts by erecting storage tanks for the water supply. A Utilities Man installs the plumbing fixtures.

The preceding example demonstrates the teamwork necessary among the various Seabee ratings to perform an operation. Teamwork is also essential between officers and men. As a Chief, you will be the link between your officers and your men, and you will be in a position to see how each person's job fits into the total effort. It is only with the teamwork of each person doing his job and contributing his portion that a battalion is able to accomplish its mission.

QUIZ

1. How can you make a group of men feel that their work is important?

2. Name three characteristics of an easy-to-follow order.

3. Under what conditions should a sup>ervi8or delegate authority?

4. When authority is delegated to a man, what should be made clear to him?

5. Name four character traits that help in getting along.with people.

6. What factors should be considered in setting up a job plan?

7. In a teaching situation, what should the instructor do when he does not know the answer to a question asked by a student?

8. Who is responsible for the safety of a work detail?

9. What Navy publication discusses safety precautions for naval personnel?

CHAPTER 2

ADMINISTRATIVE DUTIES PUniNG IT ON PAPER

As first or chief, you will find that your duties include paperwork. Station orders, work requests, requisitions, blueprints, memos, and other written material will frequently cross your desk. If you are thoroughly familiar with your field duties, the office work should not cause you any particular difficulty. However, there are a number of practices that you sliould follow and a number of procedures with which you must become familiar. Among these are requisitioning, inventorying, and surveying.

It is absolutely essential for j'^ou to be systematic in maintaining files and records. For a time, you might be able to keep your administrative responsibilities in your head; but, as work requests, reports, and publications pile up, you will find it necessary to arrange them so that each one is readily accessible when needed. A neat and orderly file pays off' when you need an old report in a hurry.

Tossing papers on your desk or filing them haphazardly in a drawer will eventually cause your records to become

hopelessly fouled up. Try to keep your files so that the man who steps into your job when you leave it can understand the status of all job assignments and work requests.

Your shops officer will judge you primarily on the speed and quality of your fieldwork, but he will also be influenced by the way you handle your paperwork. In order to perform such administrative duties as making out requisitions, taking inventories, and helping with surveys, you must understand the supply setup and the component system. You will find certain electrical publications helpful, both in your field-work and in your administrative duties.

THE SUPPLY SETUP

The Bureau of Supplies and Accounts (BuSandA), one of the seven Navy Bureaus, supervises most of the Navy's supply activities. This is big business. Millions of items must be stocked, cataloged, inventoried, and shipped to the various ships and stations. However, since the Bureau of Yards and Docks uses a tremendous amount of construction material that is not required by the other Navy Bureaus, a separate catalog of shore base equipment has been compiled. This catalog is called the bureau of yards and docks SECTION. All the items common to both ship and shore stations will be found in the regular stock catalog, which is called the general stores section. The Navy maintains special supply depots both in the States and overseas, which stock only Seabee equipment and supplies. These supply bases and centers are operated by officers of the Supply Corps. Figure 2-1 shows you a typical advanced base supply depot.

Both the BuDocks Section and the General Stores Section are divided into related types of equipment. These groups are called classes. Classes run from 1 through 99 and are arranged numerically in the catalog. Class 15, electric cable a, wire, and class 17, electrical equipment a wire COMMUNICATION APPARATUS, are of particular interest to you. Many of the items you requisition will come under one of these classes.

Figure 2-1.—Advanced base supply depof.

Suppose you want to order a growler for testing armatures. It is safe to assume that this item of electrical equipment will be found under class 17. However, you don't have to guess. Look in the general alphabetical index for GROWLER. This index shows that growlers are in class 17. Now e.xamine the index in class 17, and you will see that growlers are item number 202. Item numbers have been assigned within each class to aid in locating an article in that particular class. Item numbers run in ascending sequence from the beginning of the material listing. The first listed article is number 1, the second is number 2, and so on. Often these numbers run into the hundreds.

Remember that item numbers are merely locating num-bei*s; never use them in requisitioning or accounting for material. Use the Standard Navy Stock Number (listed underneath or opposite stock no.) for these purposes. (See fig. 2-2.)

Notice the Yl7 opposite the stock mimber in figure 2-2. The prefix Y denotes an item which is under the inventory control of the Yards and Docks Supply Office. The figure 17 means tliat the article is in class 17. A prefix YS identifies an item of restricted issue under the inventory control of the Y & D supply office. The prefix C identifies articles under the inventory control of BuDocks. In general, large none.xpend-

ARMATURE TESTER (OrowUr)

ADJUSTABLE. USED TO LOCATE SHORTS AND OPEN CIRCUITS
IN ARMATURES. 2 INCH DIAMETER AND UP. OPERATES ON
110 VOLT, 60 CYCLE CURRENT.
MARTIHDALE ELECTRIC CO. TYPE U-2 OR EQUAL.
UNIT OF ISSUE: EACH
ITEM NO. 203
STOCK NO. Y17-G-5f30
Fi3ure 2-S.—Stock catalos lilting method.

able equipment such as generators, switchboards, and power transformers, carry the C prefix. Smaller electrical equipment is usually under the inventory control of the Y & D supply office and carries the Y prefix on the stock number.

THE REQUISITION

Most of the equipment which you will require will be issued from your supply room after you submit a stub REQUISITION, NavSandA Form 307. At most bases, your company and shops officer must sign the requisition before it is filled. At some bases, you may be authorized to sign it.

There will be numerous occasions when you will need to order repair parts not in the supply room. Since all repair parts are not listed in the stock catalog, the supply officer will frequently insist that a repair p.\rts requisition be prepared in addition to your regular stub requisition. This is to prevent any possibility of a mistake because of insuflici-

cut information on your request. A repair parts requisition should always be accompanied by a stub requisition covering the same items. Always give every bit of information available about the equipment for which the repair part is desired. Name-plate data, serial number, model number, size, capacity, horsepower, name of manufacturer, etc., must be included, if you expect to receive the exact part which you desire. You may also make reference to an item on a certain page of a specific manufacturer's catalog. This method frequently proves very effective. It gives the person filling the requisition the opportunity to "see" the item you desire.

Your shops oflficer is required to hold an annual inventory of all nonexpendable equipment in his department. Nonexpendable equipment includes major items such as pumps, motors, generators, and transformers, for which an accounting must be made. The supply officer originally issues these items to the shops officer on a signed equipage stock card AND CUSTODY RECORD. He, in tum, will require you, or the equipment custodian, to sign an equipage custody record. The back of this form provides a space to inventory the items for which you have signed.

Your shops officer may also require that you take an inventory of all items within the electrical department. His purpose would be to determine that you are properly safeguarding the tools and material which have been entrusted to you. Since you are going to be held responsible for the tools and materials in your shop, keep track of them, have your men sign for the tools or keep a log of their whereabouts. Then, when the "day of reckoning" comes, you will be able to account for tliem.

THE SURVEY

Equipment is removed or junked due to wear, damage beyond repair, or simply because it is obsolete and of no further value. But you can't just take the equipment out and dump it in the GI can. It must be accounted for and examined with reference to its condition by a survey board.

The survey board is appointed by the commanding officer. For items which are valued at less than $100, the board may be composed of only one officer and is termed an informal survey board. When the value of the item is more than $100, the board is usually composed of three officers and is then termed a formal survey board. In either case, the board considers the survey request and makes a specific recommendation to the commanding officer concerning disposition of the item. If the recommendation (survey) of the board is approved by the commanding officer, the item is then disposed of in accordance with the report of the survey board. Your part in the picture consists of initiating the survey request. This request should contain full information about the piece of equipment, why it should be surveyed, where it is located, and other pertinent data.

FUNCTIONAL COMPONENTS

Once an advanced base is established and starts operating, you will follow the supply and accounting procedures just described. Setting up a base presents a special problem in supply. You will not have time to order equipment or material that is needed immediately; such material must be on hand. The functional component system is the method used by the Seabees to ensure that a newly established base has the necessary supplies to begin operations.

A functional component (generally referred to as a component) is an assembly of men and equipment, of men only, or of equipment only. A typical component contains both the men and the material necessary to perform a particular type of construction. A specific assignment might be to construct a hospital, radio station, ship repair facility, or photographic laboratory. In fact, the component system is designed to allow the immediate construction of any facility found at an advanced base. The component carries the necessary supplies and material to fulfill its mission; this makes it unnecessary to follow the comparatively slow procedure of ordering supplies by the requisition method.

ASSEMBLIES

Time is precious when construction starts on newly occupied positions. Refrigeration, power, communications, water supply, and cooking facilities must all go in to operation within a matter of hours after landing. There is no time for building permanent buildings of concrete or wood.

Simplification of construction methods has been achieved through the use of assemblies. Buildings, complete down to the last nut and bolt, are prefabricated and packed in crates. Generators are self-contained units that can be started immediately. Telephone assemblies are complete with switchboards, batteries, subsets, and cable. As with every other major piece of equipment, the necessary tools, supplies, drawings and instructions for assembly and operation are included. Sets of repair parts for the most frequent types of breakdown are included also. Placing the equipment in operation consists mainly of opening the crates, arranging the equipment, stringing the wires, tying in, and testing.

. KEEPING LOGS

Written records of events or performances listed in the sequence of their occurrence are called logs. Generator meter readings, readings of batteries on charge, and machinery operation records are types of logs. Make your log outlines simple and easy to follow so that watchstanders will have no difficulty keeping them correctly. See that the logs are always accompanied with a set of instructions.

MISCELLANEOUS OFFICE DUTIES

There are a few other duties which you must perform in the office. Here are some of the most important:

1. Keep a record of the type, size, and location of your equipment. Usually, keeping the blueprints which were used during construction will fulfill the bulk of this requirement, but, be sure to note on the blueprint all changes as they occur.

2. Tag all equipment coming into the shop for repairs. This applies especially to motors, pulleys, belts, and similar items. The same man is not always available to replace the item or return it after it has been repaired. Lack of tagging frequently causes an extra trip to isolated projects merely because two similar items were jumbled in the shop.

3. Post watch lists, fire bills, duty assignments, and other routine orders, in conspicuous places so that your meD may be fully informed as to their special details as well as regularly assigned duties.

4. Prepare estimates of material for new projects. This is a rather complex assignment and is discussed in detail in chapter 3.

PUBLICATIONS

Just because your officer gives you an assignment that is "over yoiu" head," don't throw up your hands and figure that you can't possibly handle the job. There hasn't becD anyone in the Seabees yet who knows all the answers. That's the reason why there are hundreds of publications on the many phases of electrical construction, installation, and maintenance. And also the reason why good supervisors maintain their own personal reference library. |

You should accumulate a small library of reference ma-! terial. Naturally, this library should be slanted towards your specialty such as communications, power stations, line construction, or interior wiring. But, you should also keep at least one good publication concerning each of the other phases of electricity. They will prove invaluable to you sooner or later. Generally, your publications can be divided into the following categories:

. (a) Engineering references (6) Regulations
(c) Training manuals
(d) Catalogs

A minimimi Library should consist of at least one publicatioa covering each category.

ENGINEERING REFERENCES

Engineering references axe technical publications furnishing engineering data concerning electrical theory and equipment, resistance, wire sizes, capacitance, inductance, power factor, and effects of heat. In fact, technical explanations of practically all electrical phenomena are foimd in a good engineering reference book. Such a publication will quickly supply you with many of the answers which otherwise would only be available from a collection of hundreds of regular textbooks.

There are many excellent references that are obtainable from commercial sources. One that is particularly suitable for a CE is the National Electrical Code Handbook, by Abbot, published by the McGraw-Hill Book Co., New York. This book serves both as a code (discussed below) and as an engineering reference.

Other good reference books are Lineman's Handbook, by Kurtz, pubUshed by the McGraw-Hill Co.; and Electric Motor Repair, by Rosenheim, published by Rinehart Books, Inc., New York.

ACCORDING TO CODE (REGULATIONS)

Regulations are paramount in electrical work. Improperly installed electrical equipment is dangerous, both to personnel (injuries) and to buildings (fires). Minimum requirements have been established not only for the proper installation but also for the manufacture of the equipment itself. These regtdations are called the electrical code. For the Navy or civilian electrician, they are the "Bible" of his trade.

The importance of the codes could not possibly be overemphasized. Stated simply, it amounts to this: it is IMPOSSIBLE for you to qualify as a construction electrician unless you are thoroughly familiar with the sections of the code pertaining to your electrical specialty. Power lines, communications, power plants, and interior wiring all have their regulations. You must know them before you are fully qualified for that particular field of work.

There are four main* electrical codes which are standard for the Navy:

1. Electrical Apparatus, Distributing Systems, and Weiring. Number 9Y, BuDocks, Navy Department.

2. Design Criteria jor Advanced Base Construction, BuDocks, Navy Department.

3. National Electrical Code, National Board of Fire Underwriters, 222 West Adams Street, Chicago, 111.

4. National Electrical Safety Code, Handbook H30, National Bureau of Standards, Washington, D. C.

The first code, commonly called 9Y consists mainly of specifications and rather stringent regulations that pertaiu to permanent Navy construction.

Design Criteria for Advanced Base Construction is a BuDocks publication containing basic information pertaining to construction at advanced bases. It lists the publications that apply, and discusses the types of construction that should be used at various types of advanced bases. Architectural, structural, mechanical, electrical, and sanitary requircments are covered. Under electrical requirements such subjects as selection of generators, power equipment, telephone systems, and distribution systems are discussed. One section of this publication applies to temperate zone and tropical locations, and a separate section covers Arctic locations.

Both 9Y and Design Criteria are important for Navy construction. The former is the guide for permanent construction, and the latter is the guide for temporary construction. The engineering office or the operations office of your base will probably have copies of these publications. Much of the information contained in Design Criteria is available in Advanced Base Drawings, NavDocks P-140.

Roughly speaking, the National Electrical Code covers electrical installations in and around buildings and structures, including the wires going to the power supply at the pole. The National Electrical Safety Code covers power lines, communication lines, and high voltage equipment. Since these facilities overlap in electrical usage, there will be

occasions where duplications occur. For example, there are regulations concerning switchboards in both code books. Copies of the National Electrical Code are available at the address previously mentioned. Always request the latest edition plus any supplements issued. The National Electrical Safety Code is available from the Government Printing Office. The main code book is Handbook H30. Supplements are issued covering specialized installations, but very few pertain to your type of construction work.

You are going to find numerous instances in the following chapters of statements that certain installation methods MUST be followed, that certain equipment must be used, and that the code requires such and such. All these statements are references to regulations contained in one or more of the four codes previously listed, or in United Stales Navy Safety Precautions, which was discussed in chapter 1.

TRAINING PUBLICATIONS

Perhaps the training publications with which you are most familiar are the Navy training courses prepared by the Bureau of Naval Personnel to help you prepare for advancement in your rating. These NavPers books familiarize you with the general operation, maintenance, and installation of shore-based electrical equipment, and with the military requirements for petty officers.

Publications which will prove of value to you are the instruction manuals furnished with each piece of new equipment. Every major item such as generators, switchboards, prefabricated buildings, etc., contains full instructions on the proper method of assembly, installation, operation, and maintenance. Save them! Additional copies are seldom available.

The Department of the Army also has several hundred excellent manuals partiaUy adaptable to the installation, operation, and maintenance of Navy shore-based electrical facilities. The Army manuals are called TM's (Technical Manuals) and FM's (Field Manuals). They are listed in

326605'—55 3

SR 310-20-3, Index oj Military Training Pvhlicaiions, and SR 310-20-5, Index of Military Administraiive Publicatiom. Army manuals may be available through your Information i and Educational Officer.

CATALOGS

Two or three catalogs on different types of electrical material will prove valuable to you on many occasions. Here are a few examples of the assistance they provide:

1. As references for the preparation of requisitions.

2. To increase your knowledge of electrical material. •

3. To illustrate an unfamiliar item to other personnel. Nothing helps an explanation more than a picture of what you are trying to describe.

4. To secure correct technical information concerning equipment for which repair parts are desired.

QUIZ

1. What identifying number should be placed on the requisition when equipment is ordered?

2. Who appoints a survey t)oard?

3. What system is designed to ensure that a newly established base will have the necessary supplies and equipment?

4. What BuDocks publication contains basic information pertainine to construction at advanced bases?

5. In what type of references are technical explanations of electricjl phenomena found?

6. What training publications are prepared by the Bureau of Nav»l Personnel, to help

enlisted personnel prepare for advancement in rating?

CHAPTER 3

WORKING WITH ELECTRICAL DRAWINGS FROM PLANS TO POWER

Do you want to make chief? Well, can you, working with a crew of men on a job, start with only the prints fresh from the drafting office and wind up with a completely wired building? Are you able to install a power hne or a telephone switchboard in accordance with blueprints? At least one of these abilities is essential if you are going to fulfill the requirements for Chief Construction Electrician's Mate.

It sounds complicated. But actually the installation of any electrical equipment involves the application of only three fundamental procedures:

1. The equipment is installed and connected according to the plans and specifications.

2. The equipment is installed according to the requirements of the electrical codes.

3. Standard installation methods are followed in accordance with the requirements for each job.

The codes are covered in the code books mentioned in chapter 2 and to a certain extent in other chapters of this manual. The standard installation methods are covered in other chapters of this manual and in the training manual for Construction Electrician's Mate S and 2. So for the present let's see how well you can make out on the first procedure mentioned above. You'll start with only the plans—^read them thoroughly, section by section; then you'll place and wire the equipment; and finally—if everything has been installed properly—^you can throw the switch and start the machinery humming.

The rest of this chapter describes in detail how to use a set of plans to install electricity for power and lighting in a group of quonset huts.

PLOT PLAN

The first section of the prints you are interested in is the PLOT PLAN, which gives you the whole layout at a glance. The plot plan is shown in figure 3-1, Let's check it over. Five newly constructed quonsets, 40 by 100 feet, must have electricity for both power and lighting. The heavy line across the bottom of the drawing indicates that underground ducts carrying power cable have already been installed. | MH No. 22 means manhole 22. You must tap into the j deenei^ized power cable at this manhole. Lines must be nm j from there to building 126. Most of the lines will be over- I head, and the overhead lines will dead-end at the rear of | building 126. I

NOTES j

The notation "see notes 4 & 5" at the bottom of the plot I plan refers to notes 4 and 5 listed under "General Notes" (fig. 3-2). Note 4 specifies that the crossarms are to be single except at each end of the line and at points where the line changes direction. This note also specifies the type of cross arm to be used and the insulator pin spacing. Note 5 specifies the method of guying the poles and requires that TWO strain insulators be inserted into each guy line.

Referring again to the plot plan, you will find (1) notations

Fisure 3-1 .—Plot plan.

r^arding the distance between the poles, and the size and type of conductors to be used, and (2) a reference to note 1. Checking this reference with figure 3-2 you learn that trees are to be trimmed or removed, as directed by the construction oflScer.

The plot plan and its accompanying notes probably would prove sufficient if the job consisted only in setting poles and stringing overhead lines. If it consisted only in this, standard practices and methods used in setting and guying the poles, in tying the wires to the insulators, and in dead ending the wires could be followed and the result would be a thoroughly acceptable installation. But the job includes more than this. Lines must be run underground from the manhole to the pole, up the pole to the pothead, and from the pothead to the conductors, which are strung on the crossarms. Behind building 126, lines must be run down the pole to a pothead, from the pothead through an undeiground conduit to a concrete slab, to a pothead, through a disconnect switch, then over to a transformer bank.

DETAILS

Notations on the plot plan, such as "see detail A" and "see detail B," are references to additional drawings, which furnish the detailed information required.

Figure 3-3 shows the first half of detail A. Exact dimensions are given for the location of the pole (center line of pole 8 feet from center line of manhole). Dotted lines indicate items which are not visible and which are usually underground. For example, the conduit from the manhole to the pole is represented by a dotted line, which indicates that the conduit is to be run underground. Notations along this dotted line tell you that a 4-inch steel conduit must be used between the pothead on the pole and EMH 22 (existing manhole 22). The notations also tell you that the cable pulled into the conduit should be 1-3/C #1/0 VCL.

•
8
o c
2
m
c «
04
I
«
on

Figure 3-3 gives you additional information as to t]» method of guying this pole. It specifies a %-inch stwl cable, an 8-incli (or similar) screw anchor, and strain insulators. The wrap-around method is to be used for sccuriiK the guy wire to the pole.

Figure 3-4 shows the second half of detail A. This half of detail A is an elevation of that portion of figure 3-3 whid

Figure 3-3.—First hnlf of detail A drawins.
32

is designated by the arrowB between the letters A-A. Here you will find specifications as to the type of pothead to be used, the length and class of pole, and further references to; the notes. By referring again to the notes in figure 3-2, you learn that three lightning arrestors are to be installed on this pole and on the pole at the other end of the power-line run. This provides two lightning arrestors for each of the three power lines. You also learn from the notes that the arrestors are to be grounded with bare wire connected to a rod driven into the ground at the base of the pole and that porcelain pin-type insulators and insulator pins of locust or similar wood are to be used.

BLUEPRINTS

Each of the sections of the plans mentioned so far—plot plans, notes, and details—are portions of one large blu^ printed sheet. These blueprinted sheets come in various sizes, depending mainly on whether they are to be used as flat drawing sheets or are to be rolled up. The sizes that are used are shown in figiu-e 3-5. Those sizes that are circled are the ones most frequently used by Seabee draftsmen.

Table I* Finished sheet sixes (inches)

Figure 3-5.—Drawing sheet sizes. 34

FIRST BLUEPRINT

The only item on your first blueprint that has not already been mentioned is the title block. This, as you know, contains the drawing title, its date and number, and the names of the draftsman and other persons connected with the preparation of the drawing. Figure 3-6 shows the title blocli of the first blueprint for building 126.

Fisurc 3-6.—Title block of the first blueprint.

SECOND BLUEPRINT

You should understand by now the installation of the pole line and the conduit run from the manhole to the pole. But how about the section of the installation behind building 126 where the overhead line terminates? You have several transformers, switches, and conduit runs to be installed at that point. All the necessary information covering that section will be found on your second blueprint. So, suppose you roll up the first blueprint for the time being and spread out the second one.

Do you remember the notation "see detail B" on your plot plan (fig. 3-1)? That notation referred to the section behind building 126. Drawing 472,151 is the drawing that you are now using, and figure 3-7 is detail B.

DETAIL B

There's the pole, up in the left-hand comer of figure 3-7, where the overhead power line dead ends. Notice the double crossarms, the location of the pothead, and the method of guying the pole. For this pole, a stub guy is specified.

Referring to the notes again (fig. 3-2), you will find that the guying methods for all poles are specified as those used in standard practices. Thus, you can use the same type of ground anchor for the stub pole that you used for guying the other poles. And, since you have insulators in the guy wire between the main pole and the stub pole, you won't need to use insulators in the guy wire between the stub pole and the ground anchor.

That takes care of installing and guying the pole. Next you must run an underground conduit over to the transformer bank. It's the dotted line again. It is specified as a 4-inch conduit containing 3-1/C #1/0 VCL, and it must be j placed at least 1 foot 6 inches below grade. You must place your underground conduit before the Builders pour the concrete slab on which the transformers are going to be mounted. Therefore, be sure to schedule the work of your men so that other ratings on the job are not delayed in i performing their part of the installation. The detail draw- ' ing at one side of the transformer-bank drawing shows the exact construction of the concrete slab (sec.A-A)—it gives dimensions of the slab, specifies that wire mesh is to be used , for reinforcement, and specifies that this wire mesh is to be j imbedded in the slab from one end to the other and from one ' side to the other.

See the notations on the drawing, Z" x 3" L? They refer to three angle iron posts, 3 by 3 inches in size, that will be used to support the bus run along the primaries of the transformer. Either an angle iron or another type of support must be used also to support the six cutouts. Be sure that the Builders set them in the concrete exactly where they are required. You won't have to worry about the wire fence

around the transformer bank (fig. 3-7), for the Steelworkers will erect it for you.

Are you ready now to pour the concrete? Better go back and check detail B (fig. 3-7) to make sure that you haven't overlooked anything. What are those dotted lines running from the transformer secondaries over to the building? They indicate underground conduits. One conduit is 2% inches in diameter, and the other is 3 inches. Your men must lay them in position before the slab is poured or you'll be "fouled up for fair." Don't worry about the wires now, since they shoiJd not be pulled into the conduit until the concrete has set. Just be sure to plug the ends of the conduit to keep out the concrete. The Builders can start pouring concrete now.

When the Builder says the concrete has cured, set the six large transformers. You must be extremely careful to check the location of bolt studs, whenever they are used, BEFORE the concrete is poured. Why? Because transformers should be arranged in a straight line, and it is too late to make changes once the concrete has set. Always keep this in mind—If the job is worth doing, it is worth doing right; for nothing identifies a poor electrician and a poor supervisor quicker than sloppy workmanship.

When the transformers are lined up properly, start mounting the six disconnect switches. These, of course, are mounted on the angle iron supports that are imbedded in the concrete. The bus, also using an angle iron support, is then run. The installation and connection of the required connectors, potheads, and wires is simply a matter of following standard installation procedures.

WIRING DIAGRAM

You should study one more item on your second blueprint ^fore attempting to wire the transformers. It is the wiring fcagram. The wiring diagram shown in figure 3-8 is of the ingle-line type. It gives the entire wiring system from the pet and manhole to the point where the wiring is brought 5) inside of building 126. At this point the interior wiring

Fisure 3-8.—Single-line wiring diagram.

system begins. The equipment and its sequence in the circuit, the number and size of the wires, and the transformer connections are all shown at a glance. Notice how the transformer connections are illustrated by the proper symbols. The three 25 kv.-a.'s are connected delta-delta and the three 10 kv.-a.'s are connected delta-Y. You must always carefully check the wirmg diagram before attempting to wire any type of circuit. You can gain the respect of your men only after demonstrating your ability over a long period of time, but you'll quickly lose their respect if you cause them to perform needless labor simply because you make a careless mistake.

LOOKING OVER THE LOCATION

You've seen all the blueprints for the power line from the miiihole to the inside of building 126. Do you think you can do the job now? Let's go out to the area and see what the location looks like. Figure 3-9 is a perspective of the area. Notice the manhole in the foreground of the picture. That's the same manhole which the drawing specifies as the starting point for your power line. The middle quonset—in the row of three across the open area—is building 126. The transformers will be placed at the rear left-hand corner of this building. The power line will run behind the buildings on the left and then across to the back of building 126.

LAYING OUT THE JOB

You have copies of the prints and you've seen the area. Now, it's up to you to lay out the job for your men. If you have only 3 or 4 men assigned to the job, they will work together at one location. But if you have a group of 8 to 12 men you will find them falling all over one another; coDsequently, you will need to divide the group into several crews. Then, each crew of men can go ahead with its assigned portion of a particular phase of the work. Thus, each phase will be completed in the shortest possible time. This particular installation can be laid out into four phases, as follows:

1. Setting the poles and running the overhead lines.
2. Installing the conduit from manhole to pole and from pole to transformer bank.
3. Installing the transformer bank.
4. Running the cable through the conduit and making the proper corrections.

TAKING OVER

There's your job. Draw your material; then instruct and assign your men. From now on it is strictly a question of your knowledge of (1) the code, and (2) the standard installation methods. The proper cutting, bending, and threading

39

of conduit, the stringing and tying of wires, the installing of poles and crossarms, the guying of poles, and all other installation requirements for this job now depend on your past experience and training. If your training and experience have been complete, 3'ou should be able to finish the job in short order.

TESTING

After the construction work is finished you still have two more hurdles to jump. One is testing the entire circuit; the other is getting the job inspected. Testing the circuit consists of (1) checking for grounds, shorts, and opens; and (2) testing the resistance value of the insulation, splices, insulators, and the like. These testing methods are the standard methods described in the basic texts on electricity, and in Construction Electrician^s Mate, 3 & 2.

INSPEQION

Inspection consists in getting your work approved by the officer in charge of the project. He will check the splices, transformer connections, workmanship, and very likely will insist that the insulation tests be conducted in his presence. He will require that you accompany him on his tour of inspection. Have your leading petty officer take charge of your men who are policing the area, while you inspect the job.

MANHOLE AND POLE

The first thing to be inspected is the manhole. The officer most likely will ask that the cover be removed so that he may inspect your lead-burning, which was required when you. tapped into the deenergized power cable at this point.

The conduit run from the manhole to the pole has been covered with dirt. However, the officer will carefully examine the run from the base to the top of the pole.

He will examine the pothead at the top of the conduit, the insulators on the ci-ossarms, and the lightning arresters suspended on brackets attached to the crossarms From

the top of the pole he will trace the overhead power line behind the buildings to the rear of building 126. Most of this run is visible in figure 3-10.

REAR OF BUILDING 126

In the rear of building 126 the overhead power hne dead ends. This is shown in figure 3-11. At this point the officer will also inspect lightning arrestors, crossarnis, insulators, pothead, and the conduit coming down the pole Here also

Figure 3-12.—Transformer bank.

the underground portion of the conduit run has been covered | with dirt. So the officer will next examine the conduit where | it emerges from the concrete slab. You can see this clearly in figure 3-12.

TRANSFORMER BANK

The officer is going to be very thorough in checking your , transformer-bank installation. He will inspect all the connections, potheads, disconnect switches, insulators, the primary, and the secondaries, not only for conformity with the plans and specifications but also for the quality of your workmanship. Tliis means that he is going inside that wire fence in order to make his inspection.

Figures 3-13 and 3-14 show closeups of the installation. The primary is the bus bar, and the secondary is tapped into the front of the transformers. After entering the conduit, the secondary goes underground into building 126.

Figure 3-13.—Closeup view of disconnect switches and 25-kv.-a. transformers.

Figure 3-14.—Closeup view of bus bar, lO-kv.-a. transformers, and one secondary.

How closeh' does the installation coincide with your idea of how the completed installation should appear? If you had considerable trouble making a mental picture of the installation from information on the blueprhits, you should stop right now and go back over all the drawings—plot plan, details, elevations, notes, and wiring diagram. Compare the drawings with the corresponding illustrations which sho«' exactly' how the various phases of the installation look alter completion. In order to meet the requirements for Chief Construction Electrician's Mate, you must be able to visualize a completed installation with the help of only the blueprints and specifications.

INTERIOR OF BUILDING 126

The officer also will inspect the lines inside of building 126. And when he gives his approval the disconnect switches can

be closed at the transfoniier baiik and the power cable energized from the substation to provide power and lighting for the newly constructed building. A crew working inside the building should complete the inside wiring by the time the outside work is completed.

Building 126 and the four other large buildings in the same group are 40- by 100-foot quonset huts. You're going to come upon thousands of quonsets in advanced areas. Figure 3-15 shows what a 40- by 100-foot quonset looks like inside. The special rib-type construction utilized in these buildings makes it possible to construct them in any desired length. The 100-foot length is standard for the 40-foot width.

The quonset is adaptable to hundreds of uses—machine shops, carpenter shops, automotive repair shops, electrical repair shops, and warehouses, to name only a few. Since a quonset is erected in the same way for each of these buildings, it is unnecessary to require a draftsman to redraw the plans every time somebody wants a new quonset. Therefore, only one set of erection drawings has been prepared but many copies have been printed. These copies are available to all Seabee battalions or CEC officers who have to erect the huts.

Figure 3-15.—Interior view of a 40- by 100-foot quonset.

This situation also holds true for most of the electrical, ! plumbing, and heating systems;

the partitions, the furniture j (work benches, tables, and so forth), and other portioos of j a quonset that are used in more than one structure. Almost ' all of the shops, and scores of other buildings use the same size of hut (40 by 100 feet) and the same standard electrical wiring system.

YARDS AND DOaS ELECTRICAL DRAWINGS

Since all these shops use the same general w^iring plans, electrical drawings can be made and copies of one or more of these drawings used for wiring almost all quonsets. The Bureau of Yards and Docks has prepared many stai^dard drawings. These drawings pertain to practically every type of building and facility found at advanced bases. Standard drawings are found in Advanced Base Drawings, NavDocks P-140. The following five cover the electrical installations in the 40- by 100-foot quonsets:

1. Standard Ugh ting circuit for 40- by 100-foot steel building. (Y. & D. drawing 303,665.)

2. Standard receptacle circuit for 40- by 100-foot steel building. (Y. & D. drawing 303,664.)

3. Standard 100-ampere power bus for 40- by 100-foot steel building. (Y. & D. drawing 504,130.)

4. Standard 200-ampere power bus for 40- by 100-foot steel building. (Y. & D. drawing 504,131.)

5. Standard 400-ampere power bus for 40- by 100-foot steel building. (Y. & D. drawing 504,132.)

These drawings are periodically revised and given new numbers, but the titles remain the same.

These five drawings, of course, couldn't possibly cover every conceivable situation that might arise in the field. Occasionally it is necessary for battalions to make slight revisions or additions to these drawings in order to take care of local conditions. In general, though, almost aU quonsets used for shops are wired in accordance with these five standard electrical drawings.

It is easy to visualize the great advantages that are gained through the use of standard drawings. Once a crew has erected several quonsets and become familiar with the erection details, they can cut down the time required for erection to less than a week. Crews of men, each crew performing a particular phase of the work, can move from hut to hut and erect an entire city of quonsets in less time than it takes to build an ordinary house.

YARDS AND DOCKS SHOP DRAWINGS

The Yards and Docks standard drawings are very helpful for erecting, wiring, heating, and installing the plumbing in almost all quonsets. But how about the arrangement of the machinery in the building? A drawing for a carpenter shop doesn't work so well for a machine shop, since the machinery in the two shops is vastly different, so it is necessary that at least one drawing for each of the different shops be prepared. These drawings serve four important purposes by furnishing the following information:

1. A FLOOR PLAN, which shows the layout and arrangement of machinery, equipment, and space requirements for a particular shop.

2. A BILL OF MATERIAL, which lists all the machinery, equipment, and material required to erect, furnish, and make ready for operation a particular shop.

3. A list of REFERENCE DRAWINGS (Y. & D, standard drawings), which are to be used to erect, wire, and install the heating and plumbing in a particular shop.

4. Notations about any electrical, heating, plumbing, or erection requirements in addition

to, or at variance with the standard drawings covering each of these phases of the construction.

ADVANCED BASE MACHINE SHOP DRAWING

Let's take a look at figure 3-16, which is one of the Yards and Docks advanced base machine shop drawings. This

drawing contains all four of the items mentioned above—^laj out of the equipment, a bill of material, reference drawing! and notations concerning the particular installation. Yw can't teU much about this drawing from figure 3-16; so sup pose we blow it up and take a look at it section by section.

FLOOR PLAN AND NOTATIONS

The floor plan illustrated in flgure 3-16 is shown again ii figure 3-17. You should be able to distinguish some of thi notations about ihe various portions of the instaUation. 0; course, the electrical notations are of chief concern to you Four switches—one master switch of 200 amperes and thre< small switches of 30 amperes—are to be mounted in the uppef left-hand corner. The locations of the receptacles along th« walls are indicated. The four rows of circles, which represent the overhead droplights, indicate that 32 droplights are to be installed at equally spaced intervals. The location of the power bus is indicated in a general manner. Variations of this drawing, to meet local conditions, may include (1) placing the switches in another comer of the building so as to simplify the connections to the nearest pole, (2) increasing the lighting at one or more benches, or (3) other similar changes. Generally speaking, the job is installed almost exactly in accordance with recommended drawings.

BILL OF MATERIAL

The machine shop drawing has two bills of material, which are shown in figures 3-18 and 3-19. The division into two lists won't mean much to you, since it merely shows which Navy biu-eau supplies the items for this shop assembly. Your main interest will be in seeing that all the equipment and material listed are accounted for. If it is, you'll be too busy getting it erected to worry about the vast naval organization responsible for having it at that exact spot the moment it is needed.

O
PLAN
i« advonccd-bflM inachin«-«hop drawins*

Figure 3-18.—Machine shop bill of material list.

REFERENCE DRAWINGS

Now look at figure 3-20. It shows the title block and reference drawings from the machine sliop drawing in figure 3—16. As you can readily see, the building is to be 40 by 100 feet. The reference dramngs are standard drawings used in the erection of almost all quonsets, and every one of them would be used in the erection of this particular machine shop. Though a considerable number of drawings are listed here, the ones that you are interested in mainly pertain to the electrical installation. For the macliine sliop, tliree standard electrical drawings are required:

1. Y. & D. drawing 303,664 (receptacle circuit).
2. Y. & D. drawing 303,665 (lighting circuit).
3. Y. & D. drawing 504,131 (formerly 303,659) (200-ampere power bus).

BILL OF MATERIAL
NOTlt

Figure 3-19.—Machine shop bill of material list.

WIRING THE QUONSET

The wiring of quoiiset huts doesn't appear to be very compHcated, does it? And it isn't after you'vk wired two

Figure 3-80.—Title block and reference drawings.

OR THREE. From then on, you'll find them all much the same. You have a chance right now to sec how one is wired, for building 126 is a quonset—it is to be a machine shop like the one in figure 3-16. Let's see how the wiring job for this building is handled.

PLANS AND MATERIAL

Getting a copy of the reference drawings (fig. 3-20) and locating the electrical material is the first job to perform. This isn't too hard, since copies of all the drawings are included in the shipment of material for the machine shop assembly. Even if some of the drawings are missing, other copies are available from the engineering office. Each reference drawing contains a bill of material showing all items necessary to install that particular portion of the wiring.

Check to see that all of the required electrical items are ii eluded in the shipment. As soon as all the prints and mf terial are assembled, you and the crew are ready to begi work. The reference drawings in figures 3-21, 3-22, an 3-23 show a typical wiring job for a quonset hut.

LIGHTING CIRCUIT

Figure 3-21 shows the standard lighting circuit drawinj (Y. & D. drawing 303,665). You may not be able to ge much help from this small illustration. So let's enlarge i' and examine it section by section. You need to be familial with figure 3-21 in order to install the lights in the machini shop. You already know the location of the lights from th€ machine shop drawing. Figure 3-23 is a picture of the details. That sheet-metal plate (item 8) is nailed to the expandable groove in the center of the rib of the building, and the lamp socket (item 18) is mounted on the plate by means of sheet-metal screws. The droplight is then assembled, as shown in the droplight detail. Between drop-lights the No. 14 nonmetallic-sheathed cable is clamped to the ribs, as shown in the "clamping cable detail." Each of the item numbers given in the detail drawings, corresponds to an item number listed in the bill of material. The items in the bill of material are described fully. By referring to it you will avoid any confusion as to where to use similar items.

RECEPTACLE CIRCUIT

The standard receptacle circuit is shown in figure 3-22. Since you may have difficulty in getting much help from the small illustration and since you already know the location of the various receptacles, let's look at the detail drawings. These are shown in figure 3-24. Notice that the method of mounting the plate (item 6), of attaching the duplex receptacle to the plate, and of clamping the cable to the wall is exactly the same as was used for installing the drophghts..

drawins for o 40- by 100-(oot qwont«t.

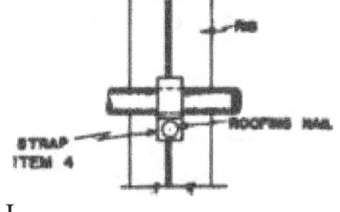

I

METHOD OF CLAMPING CABLE
TO Rib
MHON HOLCS KR^TYPC**
iGa. Fixtt OcTMU Item 6
tENCE NGS

APPROVED BV BUREAU OF DATE Wr .
FUNCTIONAL COMPONENT
irza^ > mTwf row ■utlct ■wlwi mwip.
lUVY
Advanced bases
STANDARD Receptacle Circuit a
FOR A 40'X100'STEEL
Building

m* T. ft D.
303,664

DROP LIGHT DETAIL SCALE: 1/2 sue
Figure 3-23.—Details from the standard lighting circuit drawing.

POWER BUS

You'll find the third drawing—the 200-ampere power bus—in figure 3-25. You should be able to make out most of the detail drawings from this illustration. Notice that clamps and brackets, rather than nails and screws, are used in mounting the ribs of the building. This method is necessary because th^ combined weight of the wires and the rack is too great to rely on the same methods by which droplights and receptacles are held.

THE BUTLER BUILDING

Are you wondering why a few details are shown on the drawings besides those already mentioned? The additional

details are to be used if the building is a butler building.

The details are practically the same for both quonset and Butler building except for the method of mounting receptacles, sockets, and racks to the walls and ceiling beams.

Fi3ure 3-26.—Inside and oufside views of a Butler buildins.

For mounting in the Butler building, adapters must be used. Both inside and outside views of a Butler building are shown in figure 3-26.

MISSING LINKS

You're probably wondering about several missing links between the drawings and the construction. The power available at the transformer bank outside building 126 normally isn't necessary to run a small machine shop like

the one shown in figure 3-16. However, several other buildings are in tliis area. Later, they will be furnished power and lighting from this same transformer bank. The incoming wires to building 126 are also larger than required for a machine shop of this size. They were purposely made oversize to take care of extra equipment to be used in this building later. Because of this extra equipment the wires were brought through the side of the building rather than through the end, which is the usual practice.

QUIZ

1. What steps are involved in testing the circuits?

2. What precautions should be taken to keep freshly poured concrete out of conduits?

3. Will individual drawings be made for the wiring of a lighting circuit in a quonset hut designed as living quarters? Why?

4. What three fundamental procedures are involved in the installation of electrical equipment?

6. When installing an electrical system, what section of the blueprint would you refer to first?

6. What part of a blueprint contains the drawing title, its date, and its number?

7. What diagram should be checked before any transformer connection is made?

8. How many standard electrical drawings are required for a machine shop?

9. What does the bill of material on each reference drawing show?

10. In what type of building are adapters used to mount receptacles, sockets, and racks?

CHAPTER 4

ELECRICAL REPAIRS THE ELECTRICAL REPAIR SHOP

Electrical equipment requires careful maintenance to keep it in operating order. You will need to make frequent tours of inspection to examine motors, generators, and other electrical apparatus. The electrical repair shop will be your headquarters for maintenance and repair.

In setting up a shop, you will need to do very little of the actual planning. The equipment needed in the shop is determined by the functional component system (discussed in chapter 2), and the shop layout is described in Advanced Base Drawings, NavDocks P-140. The layout will depend upon the size and the mission of the base to which you are attached. Sometimes the electrical shop is combined with the carpenter shop or the automotive repair shop.

Standard drawings 433,454; 482,424; 482,715; and 523,243 give the recommended laj'out and equipment for various types and sizes of electrical repair shops. These drawings are revised periodically to meet changing needs. If necessary, new drawings are prepared. Advanced base drawings are intended only as a guide; your commanding ofl&cer may order any necessary changes.

TOOLS AND EQUIPMENT

Although electrical repair shops vary in size and mission, certain equipment is common to every shop. Electrical repair inevitably Involves rewinding armatures and field coils and perhaps stators; consequently, tools and equipment to perform these operations must be in every repair shop.

326605'—56 5

The following tools and equipment are commonl3- used i electrical repair:

1. Bench lathe

2. Growler

3. Coil winder

4. Armature stand

5. Band saw

6. Insulation shear

7. Insulation folder

8. Test switchboard i

9. Electric oven •

10. M^^er

11. Dipping tank

The switchboard is designed to test resistance, voltage, or current of a-c and d-c machines. The electric oven is used to dry moisture out of insulation and to dry varnish on armatures and stators that have been dipped. An electric oven is standard equipment for every electrical shop. In an emergency, you can construct an oven for drying equipment by securing heat-insulating panels to suitable frames or concrete blocks lined with insulation and by supplying heat with an electric heater, a radiator, a steam coil, or a hot air furnace.

The equipment listed above is only a small portion of the equipment necessary in a shop. In addition to the small tools commonly found in a machine shop, you will have workbenches, drills, grinders, and perhaps an arbor press, a tension rack, a taping machine, a vacuum and air pump, a vacuum impregnation tank, hydraulic bender, oxyacetylene welding equipment, a pipe threading machine (for conduit), a chain hoist with hook, a portable pump, a portable generator, and a portable blower.

PREVENTIVE MAINTENANCE

A regular schedule of maintenance and inspection will add years to the life of electrical equipment at your base and will save you much repair work in the electrical shop. You can work out a general schedule of inspection for motors and

generators from the time intervals given in the accompanying chart. These intervals are intended only as a guide. Such factors as the type of installation, operating duty cycle, local atmospheric conditions, and specific recommendations of a manufacturer are equally impK)rtant in establishing a schedule of inspection. The goal of any inspection schedule is to keep the station's equipment in good condition with -minimum effort. Tables are on pages 60 and 61. - In maintenance and inspection procedures, there are two general principles to follow:

1. New equipment should be carefully watched imtil extended operation shows that it is performing satisfactorily.

2. Old equipment requires more frequent cleaning and inspecting than new equipment of the same type.

GENERAL MAINTENANCE OF MOTORS AND
GENERATORS

In general, successful motor and generator maintenance involves two rules:

1. Keep the equipment clean.

2. Keep the equipment lubricated.

The rules of maintenance apply both to large and to small pieces of equipment.

The importance of keepir^ equipment clean, both externally and internally, cannot be overemphasized. Dust, dirt, and foreign matter (carbon, copper, mica, etc.) tend to block ventilation ducts, thus causing overheating. If the particles act as conductors or form a conducting paste by the absorption of oil or water, the windings may eventually be short-circuited or grounded. Abrasive particles may puncture the winding insulation. Iron dust is particularly harmful since magnetic pulsations tend to stir up the dust and distribute it to all parts oi the machine.

Wiping is one method you can use to remove loose dust and particles. Of course, it can only be employed when the dust and foreign particles are located in accessible parts of

&

s

a) >
C
fl 3 O
t 3
02
a o
t3
o u
e
«
Is
I
•c
s I
S o
2 «
3 OS
a
u
at
T3 C
as
2
o
s
E
o Q
03
I

c
73
a
e3
to .£3
1
9
s a
o O
a
I
ss
e o
u « C
c o u

the machine. You'll find that clean dry cheesecloth makes a fine wiping rag since it will not deposit lint. Be sure that the man doing the job doesn't n^lect such parts as the end winding, mica cone extensions at the commutator of d-c machines, slip ring insulation, terminal and terminal insulation, connecting leads, etc.

The use of compressed air is another cleaning method. It is effective in removing dry loose dust and foreign particles that are located in bard-to-get-at spn^ts. Air pressure up to 30 pounds per square inch may be used on motors or generators of 50 horsepower or 50 kilowatts or less. Pressures up to 75 pounds per square inch may be used to blow out machines which are over 50 horsepower or 50 kilowatts. To be sure that only dry compressed air is used, always allow any accumulation of water in the air hose to be thoroughly blown out before turning the air blast on the machine. In addition, take the precaution of opening both ends of the machine so as to allow a path of escape for the air and dust.

The use of suction is preferred to the use of compressed air, especially if abrasive particles are present. With compressed air you always stand the chance of driving the particles into the insulation and puncturing it. Suction is particularly useful in drawing away the loose particles produced during the stoning of commutators or the seating of brushes If a regular vacuum cleaner is not available, you can improvise by attaching a flexible tube to the suction side of a portable blower. As a safety precaution, remember that hose tips for either suction or pressure should not be of metal.

The use of solvents is the fourth and final cleaning method. Whenever possible, the use of solvents for cleaning purposes should be avoided. However, the presence of hard deposits of dirt or oil will sometimes make it mandatory to use a solvent. Carbon tetrachloride was the solvent formerly used, but its corrosive effect and toxic properties are so great that its use is no longer recommended. Stoddard solution, a dry-cleaning solvent sold under the trade name of Varsol, is a satisfactory solvent for cleaning electrical equipment.

The best procedure is to moisten a lintless cloth with the fluid and Hghtly rub the sufrace to be cleaned. The dirt or oil wUl' be loosened so that it can be wiped off with a cloth or, if necessary, removed with a wooden scraper or fiber brush. After the surface has been cleaned, it should be thoroughly dried by wiping with a clean cloth and, if available, with compressed air. Stoddard solution is a safety-type solvent in which fire and health hazards have been minimized. Nevertheless, normal precautions against inhalation and against fire hazards should be observed when this solvent is used.

In an emergency, clean, fresh water may be used instead of a solvent. The water should be warm or hot to obtain best results. A neutral detergent may be added to aid in removing grease. Equipment which has been cleaned by water must be thoroughly dried before being placed in service.

The second important rule for successful motor and generator maintenance is to keep it lubricated. When an inspection of sleeve bearings reveals a deficiency of oil in the housing, oil should be added immediately. Sleeve bearings having an overflow gage should be filled until the oil is about K« inch from the top of the gage. If the machine is equipped with an oil fiUer gage, the gage should be filled to the oil level mark. Bearings that have neither an overflow gage nor an oil filler gage should be filled to a level where the oil ring dips into oil to a depth of about half the shaft diameter. Machines equipped with waste-packed bearings should be re-oiled every 1,000 hours of operation. Bearing oil should be renewed annually unless the manufacturer reconmiends otherwise. This renewal is accomplished by first removing the drain plug, draining off the old oil, flushing the bearing with clean oil until the drained oil flows free, and, finally, replacing the plug and filling with the grade of oil recommended by the manufacturer. For safety's sake never add oil to a machine that is running. This practice may cause oil to be sprayed onto the windings and to slop over the floor, causing a fire and accident hazard.

The frequency of lubrication of ball bearings depends upon such things as the service of the machine, the tightness of the housing seal, and the type of bearing. Ordinarily the addition of grease will not be necessary more often than once in 6 months. Sealed bearings usually need no additional grease for a period of about 2 years or longer. Be sure to follow the manufacturer's recommendation as to the grade and type of grease to use. And never overlubricate. Too much lubrication can cause heating and other bad effects.

BRUSH MAINTENANCE (NEUTRAL POSITION)

Direct current machines operate best when their brushes are set on the commutator at the neutral point, where the armature current reverses direction. At this point, the best conunutation is usually maintained. In most instances you will find that the brushes have been factory set at the neutral position. However, under certain conditions it may be necessary to reset the brushes; brush shifting should be done with the machine disconnected from the source of power. To aid you in this task you will find certain markings on the armature of the machine. In the lap wound armature shown in figure 4-1 (a), there are two commutator risers whose slots are marked with paint. The distance between these two marked risers is equal to the distance between the coil sides of one armature coil. Now, notice the group of three commutator bars that are stamped with the letters TT and BB. The commutator bars marked with identical letters are each connected to one of the two ends of the same armature coil. The top view of the armature in figure 4-1 (b) will help to clarify this. Since one end of each of the armature coils is connected to the same conmiutator bar, that bar will have both a T and B stamped on it. This middle bar is the mechanical neutral bar, and is the one over which the brush should be placed once the armature has been rotated into its correct position. To determine this correct position:

1. Turn the armature over until the marked riser slots are directly xmder the center lines of two commutating poles (fig. 4-1 (a)). Where there are no commutating poles, turn the armature until the marked slots are midway between two main poles, and equal distances from them.

2. With the armature set in the mechanical neutral position shift the brushes until they are centered over the middle commutator bar of the stamped group.

Figure 4^1.—Locating the mcclianical neutral.

If an excessive amount of sparking occurs on starting a noncommutating pole machine, the trouble may be that the brushes are not on the correct neutral position. Brushes of generators should be shifted slightly forward in the direction of rotation. Brushes of motors should be shifted backward. In either case, the correct neutral position is that which produces the minimum amount of sparking.

BRUSH MAINTENANCE (SPACING)

Locating the neutral point for one set of brushes will not ensure perfect commutation for the other sets of brushes placed around the commutator. This can only be accomplished if (1) the brush holders are all the same distance away from the commutator, and (2) the brush groups are equally spaced around the commutator.

Setting of brush foldei-s with relation to the commutator varies slightly with brush holder type. Usually, this distance will not be greater than ^ inch or less than Me inch. A gage of proper thickness may be improvised from strips of fiber. To set a brush holder, place the gage under the holder, loosen the setscrews or clamping bolt, move the holder down until it rests firmly on the fiber gage, then securely fix the holder in place. Repeat this operation on all holders.

Each brush group should be equally spaced around the circumference of the coninuitator. This can be checked accurately by first wrapping a strip of paper tighth'^ around the commutator and marking it at the point where the paper

laps. This step is shown in figure 4-2. Remove the paper from the commutator, cut at the lap, and fold or mark the paper into as many equal parts as there are brush studs. Kext, replace the paper around the commutator and secure the ends with glue or scotch tape. Finally, adjust the brush holders up or down slightly so that the toes of the brushes are at the creases or marks. When a set of brushes is too far out of location, it may be necessary to correct the stud position. In addition, all brushes of one group should line up with each other. This can be checked by seeing if the toes of each brush in the group line up with the edge of the same commutator bar.

BRUSH MAINTENANCE (STAGGERING)

A final check of brush position should be made to see that the brushes are properly staggered. The correct staggering of brushes is shown in figure i-3A. The brush holders are set on pairs of studs so that the brushes on one pair ride on the space between the brushes in an adjacent pair. This distributes brush wear evenly across the commutator. Brushes of the same polarity should not track each other, as at figure 4-3B, because wear on the commutator when current flows from brushes to commutator differs from wear when current flows from conunutator to brushes. For direct current machines having a niunber of poles equal to twice an odd niunber, it will obviously not be possible to stagger all the brushes in accordance with the correct method. When this condition exists, stagger all but the odd pair of positive and negative brushes in accordance with the correct method, and the odd pair in accordance with the incorrect method.

BRUSH MAINTENANCE (FIHING)

When inspection reveals that a brush has been worn down almost to the brush rivet, or that a heel or toe is chipped, the brush should be replaced. New brushes should always be fitted to the contdur of the commutator. There are two accepted methods of seating the brushes. In, one method, sandpaper serves as an abrasive; in the other method, the abrasive is a brush seater.

Figure 4-4 shows the four steps involved in installHng brushes on d-c machines when sandpaper is used as an abrasive. Before beginning, disconnect all power, and be sure that the machine cannot be started accidentally. Step one is concerned with a preliminary inspection of the brush holders. Use a flashlight to check the brush holders, inside and out, for burned spots that may have been caused by flashing at the commutator. If the inside of the holders shows roughness, remove this with fine sandpaper glued to

Figur* 4-4.—Installing brushes on d-c machines.

a flat piece of wood. Step two involves inspection of the pigtail connections of the new brushes to make sure that they are properlj' secured. If the pigtail connections are defective, current will flow from the sides of the brushes into the holdei-s and cause side burning, wear, overheating, and improper brush operation. Step three concerns the fit of the brushes in their holders. Always make sure that the brushes can move freely. In this respect, they should not be so loose that there is an appreciable side play, or so .tight that they may expand and stick at operating temperatures. In step four, the brushes are seated to the commutator one at a time. To do this, lift the brush to be sanded and insert a strip of medium sandpaper (No. 1), sanded side up, between the face of the brush and the commutator surface. Keep the sandpaper close to the surface of the commutator, and with the brush held down by normal spring pressure, pull the sandpaper in the direction of fonvard rotation of the armature. Each time the sandpaper is pulled, lift the brush and return the sandpaper to its original position for another

pull. Continue until the brush face makes a good fit with the commutator curvature and then finish off with fine sandpaper (No. 0). If a vacuum cleaner is available, use it to remove the dust while sanding. Otherwise, use compressed air after the job is done, blowing out the dust from the back toward the commutator. Never use emery

PAPER, EMERY CLOTH, OR CARBORUNDUM tO fit brUSheS 68

the particles will imbed in the copper segments of the conmiutator or in brushes.

A brush seater is sometimes used in place of sandpapei. It consists of mildly abrasive material loosely bonded into a stick about 5 inches long. The following steps are involved:

1. Install the brushes in the brush holders taking care to make the preliminary inspections shown in steps one, two, and three of figure 4-4. Start the machine.

2. Press a brush securely against the commutator by using a stick of insulating material or by increasing the brush spring tension to its maximum value. Touch the brush seater lightly to the commutator exactly at the heel of the brush so that abrasive material torn from the brush seater will be carried under the brush. If the seater is placed even % inch away from the heel of the brush, only a small part of the abrasive passes undei the brush.

3. Repeat step 2 for each brush, applying the seater for a second or two, depending on brush size. If white dust is plainly visible after seating a brush, insuflScient pressure has been applied to the brush, or the brush seater has been applied too heavily or too far from the brush.

4. After seating aU brushes, blow out the machine with dry compressed air or clean with a vacuum cleaner. Adjust spring tension of each brush to the correct operating pressure.

COMMUTATOR MAINTENANCE (CLEANING)

The action of the brushes on the conmiutator surface will produce a stable copper oxide-carbon film. Such a film will

be a uniform color which varies all the way from copper to very dark brown. A commutator in good operating order is usually a shade of chocolate brown. This uniform glaze is an essential requirement for good commutation and must be protected against the action of certain foreign atmospheric conditions. Fumes and oil vapor, for example, tend to smudge, gum, and discolor the commutator. Chlorine usually causes a green discoloration, while sulphuric fumes usually produce a blue surface. Gases, oil and grease will create a mottled appearance on the commutator surface film. You can combat these enemies of good commutation by seeing to it that the commutator is .cleaned at regular intervals.

One of the most effective ways of cleaning a commutator is to apply a canvas wiper while the machine is running. A suitable wiper can be made by wrapping several layers of hard-woven canvas over the end of a strong, pliable wood stick and securing the canvas with rivets. The strip should be long enough so that it can be held securely in both hands, about \'7di inch to ⅝ inch thick, and of width nearly equal to the width of the commutator. To prevent the possibility of the rivets coming in contact with the commutator, they should be covered with strips of linen tape. Either the face or end of the wiper can be applied to the commutator. When the outer layer of canvas becomes dirty, it can easily be cut off to expose the next layer.

The commutator can also be cleaned when the machine is secured. A toothbrush serves as an effective tool for cleaning out the commutator slots. Then, the commutator can be polished with a clean canvas cloth or cheesecloth. There are some "don'ts" to observe:

1. Don't use cotton waste or any cloth that leaves lint.

2. Don't use dry cleaning solvent for routine cleaning since it tends to dissolve the uniform glaze on the commutator surface. Where stubborn dirt makes it necessary to use Stoddard solution, apply it by the dampened cloth method. Take care that the solution does not get on the brushes.

3. Don't use a lubricant on the commutator during or after the poUshing and cleaning period. Oil and lubricants tend to develop a high resistance film, dissolve the binder out of the mica insulation, and cause carbonization of the mica. *

COMMUTATOR MAINTENANCE (TRUING)

Any condition of the commutator that will affect its balance (such as flat spots, grooves, scoring, out-of-round) must be corrected. For the most satisfactory operation, the variation of distance from the center of the commutator to its surface (the radius) should not be in excess of 0.002 inch (2 mils). This can be checked with a dial test indicator.

The easiest method of removing flat spots, grooves, and deep scratches is hand stoning. The dressing stones used in this process are merely rectangular blocks of abrasive material ranging from very fine to a coarse grade. You'll want to use the finest grade of stone that will do the job in a reasonable time. Generally, coarse stones should be avoided as they tend to produce scratches which are bard to remove. The stone should be formed or worn to the ciunra-ture of the commutator and should have a surface larger than the largest flat spot to be removed. Figure 4-5 shows the correct method of applying the hand stone to the commutator. Have the machine nmning at, or sHghtly below, rated speed. If you're working on a generator be sure to Uft all the brushes away from the commutator—the prime mover will supply the turning power. For motors, remove all but enough brushes to keep the armature turning at the proper speed—unless you can rig up some sort of external driving source. Where it is necessary to employ the motor's brushes, use old brushes that can be discarded after the stoning is completed. In applying the stone, try to maintain an even surface by moving the stone very slowly back and forth parallel to the axis of the conmiutator. Don't press too hard on the stone, just enough to keep it cutting. And keep your mind on the job to avoid electric shock, or

jamming the stone between fixed and moving parts of the machine. After stonmg, smooth the commutator surface with a very fine grade of sandpaper and then give it a poHsh by using the back of the paper. When this conditioning job is completed, remove all traces of dust, reinstall and reseat the brushes, and, finally, clean the commutator again.

Figure 4-5.—Hand stonins fhc commutator.

When the commutator is too rough or eccentric to be stoned do^vn successfully by hand, you will have to resort to grinding with a rigidly supported stone. In this process the stone is mounted on a special rest on a brush holder stud (fig. 4-6). The motor or generator is run at normal speed by means similar to those described under hand stoning operations.

It is often more practicable to remove the armature from the machine and grind it in a lathe than to perform this operation while the armature is in the machine. Whether

326605'--.IS 6 73

Figure 4-6.—Grinding with a rigidly supported stone.

the work is done with the armature in a lathe or in the nia-cliine, take great care to ahne the supports so that the motion of the stone is parallel to the axis of the machine. Failure to aline the supports will produce a taper on the commutator.

When extreme commutator roughness exists, the use of a dressuig stone may be insufficient. Then you'll have to take the armature to the repau- shop where the commutator can he turned in a lathe.

TURNING THE COMMUTATOR

The commutator should be turned down in a lathe only when stoning or grinding is insufficient to remove excessive roughness or distortion. Before turning, be certain that the armature is otherwise in good repan- and is approximately balanced. Place masking tape around the commutator risers to prevent any loose copper from getting down inside the coil leads between the armature wiring and the commutator. Support the armature in the lathe and use a

diamond point cutting tool. The cuts should overlap and not leave a rough thread on the commutator. The usual cutting speed is about 100 feet per minute. The feed should be about 0.01 inch per revolution. The depth of the cut should not be more than 0.01 inch. A heavy cut may cause the commutator to be ground into a noncylindrical shape because of elasticity in the support. In addition, a heavy cut may cause the turning tool to twist the commutator bars and to cut deeper at one end than at the other. It is therefore preferable to take a light cut, even though several cuts may be necessary.

A-C COLLEaOR RINGS AND BRUSH MAINTENANCE

The collector rings and brushes of the alternators and wound-rotor motors in the station are subject to the same wear and tear as the commutator and brushes of the direct-current machines. It will be up to you, then, to see that the a-c machines are given the same careful attention as the d-c machines. Make certain that—

1. Each brush is inspected regularly for free play in the holders and that the full surface is bearing on the ring.

2. The brush rigging is kept free from dust, oil, lint, metal particles, and dirt.

3. The collector rings are kept free from coating and scale. Suitable stiff brushes can be used to remove the major part of the accumulation. Stoddard solution applied with a cloth should follow.

4. Each generator which normaUy is shut down is run for a short time every day. This is the best way to prevent corrosion, electrolytic action, pitting, and flat spots on the rings.

Current flow during the normal operation of a generator sometimes sets up an electrolytic action that causes pitting of the surface of the rings. The appearance of this pitting is different from the pitting which results when a generator is secured, in that it is general over the whole ring area

and does not cause localized flat spots. Sometimes it is evident in one ring only. This pitting can be prevented by revereing the polarity of the rings every few days. Leads to the collector brushes or at the switchboard should be made long enough to permit easy interchanging. The phase rotation of the generator will in no way be affected by this reversal of polarity.

MOTOR TROUBLES (A-C AND D-Q

The repair of electric motors is one of the main jobs performed in an electrical shop. There are many causes of motor troubles. In many cases, the remedy is obvious as soon as the cause of the trouble is definitely known. You will frequently be able to effect a repair by merely tightening a connection, adding oil, or loosening a belt. Such repairs may be performed on the spot without taking the motor to the shop.

Other repairs are complex operations requiring the winding of armatures, field coils, or stators. Motors requiring this type of work should be tagged for shop repairs.

The following are common types of troubles in motors:
1. Motor fails to start
2. Motor vibrates or makes excessive noise
3. Bearings overheat
4. Windings overheat
5. Motor burns out

FAILURE OF MOTOR TO START

Use the following steps to determine why a motor fails to start:

1. Check the directions on the controller to see if you have followed instructions. Make certain that the switch or circuit breaker is closed and that no fuses are blown.

2. See if the overload relay on the circuit breaker in the supply line trips. If it does trip, inspect for a short circuit.

3. Check the voltage shown on the motor nameplate to see that it agrees with the voltage of the supply line. Then measure the voltage at the motor terminals with motor connected to see if the voltage drop in the line is excessive.

4. Check for incorrect connections, loose connections, open connections, or open circuit in the controller or the wiring to the motor. Use the wiring diagram to make certain that the motor is connected for the correct direction of rotation.

5. Check the motor windings for open circuit, short circuit, or grounds.

6. Reduce the load on the driven machine to see if overload is the cause of trouble.

7. Check for excessive friction due to belt tension, gear side thrust, misalinement, stiff grease in ball bearings, insufficient lubrication of sleeve bearings, bent or sprung shaft, or the rotor's rubbing the stator. If necessary, disconnect the motor from the driven machine to localize the source of trouble. If the motor starts and runs when disconnected, check the driven machine for locked or jammed parts.

8. Check for stiff or frozen bearings. Free stiff bearings. Replace frozen bearings, and for sleeve bearings, resurface and polish the shaft.

9. If an electric brake is installed, check its operation.

EXCESSIVE NOISE OR VIBRATION

All motors have a characteristic hum. Hum, itself is not objectionable, but an unusually loud hum indicates trouble which should be investigated before the defect becomes serious. Correction of the trouble may require nothing more than tightening a bolt or a loose connection.

If there is dirt in the air gap or objects are caught between the fan and the end shields, clean out the motor to remove the dirt or obstruction.

If the motor or driven machine is unbalanced, vibration may result. Disconnect the motor and determine where the trouble lies. Balance the unbalanced unit.

Bearing troubles loading to noise and vibration include a deficient oil supply, ovei^cased ball bearings, excessive end play, and too tight or too loose bearings. Add oil to correct a deficiency. Remove the drain plug to allow excess grease to run out. Change the belt alinement to remove excessive end play. Replace bearings that do not fit properly.

OVERHEATED BEARINGS

If sleeve or ball bearings overheat, check the end shields to see that they fit squarely and are properly tightened. Reduce the belt tension or gear side thrust if necessary. See that the gear side thrust is not transferred to the motor.

MisaUned couplings or belts cause overheating. Check and correct the alinement.

If the shaft is bent or sprung, straighten or replace it. Occasionally you may have to replace the entire rotor.

If too much heat is conducted to the bearings from overheated windings, correct the causes of the overheated windings. (Discussed later.)

Sleeve bearings become overheated when dirt or foreign matter gets into the oil, when the oil rings become defective, or when the bearings are too tight. Drain dirty oil and clean the reservoir before refilling. Replace defective rings with new rings, and loosen bearings which are too tight.

Ball bearings become overheated from improper alinement of the bearings, too much grease, not enough grease, or the wrong grade of grease.

To check for proper alinement, examine the bearing assembly and make certain that the races are perpendicular to the shaft. Improper greasing procedures are a frequent cause of trouble in electric motors equipped with ball bearings. Follow the manufacturer's instructions closely when you change grease; an excess of grease in the bearing housing causes grease to be forced through the bearing housing seals and onto the windings. The excessive quantity of grease in the housing causes high pressure, churning, excess temperature, rapid deterioration of the grease^ and ultimate destruction of the bearing.

OVERHEATED WINDINGS

The principal causes of overheated windings and the corrective procedures are as follows:

1. Heat conducted to the windings from overheated bearings. Correct the cause of overheating in the bearings.

2. Short-circuited coils. Locate and remedy the short circuit.

3. Overload. Check the electric power input into the motor. If possible, reduce the load.

4. Incorrect connection of motor internally or to external circuit. Check connections with the wiring diagram.

5. Rotor rubbing the stator. Check air gaps and center the rotor.

6. Restricted ventilation. See if fans and baffles are correctly assembled. Clean the air passages and the windings.

7. Too frequent starting or running at a more severe duty cycle than the motor can accommodate. If motor is on full automatic control, check operation of the control devices. Check the duty cycle with the nameplate rating on the motor.

8. Low voltage on the supply line. Check voltage at the motor terminals while the motor is running.

MOTOR BURN OUT

A motor may bum out as a result of overheating in either the bearings or the windings. An insulation failure may also cause a motor to burn out. Insulation failures may be due to excessive moisture in the windings; grease, oil, or dirt in the windings; or faulty insulation. If possible, a burned out motor should be repaired with the repair parts provided. If extensive, repairs are required or repair parts are not available, you will have to replace the motor.

A-C MOTOR TROUBLES

In addition to the troubles common to both a-c and d-c motors, there are troubles peculiar to each type. In a three-

phase induction motor, for example, failure of the motor to start may result from the motor's having become single phased. In such a case, check the power line to be sure that power is available on all three phases. Also check connections from the supply line to the motor for open circuits. The motor winding should also be checked for open circuit.

A humming noise may indicate that the three-phase motor is running single phase. Stop the motor and try to restart it. If single phased it will not start, and you should correct it in the manner described in the preceding paragraph.. A humming noise may also indicate that the currents in the three phases are unbalanced. Measure the current; if they are unequal, check the windings for open circuits, short circuits, and grounds.

An overheated three-phase a-c motor may result from running single phase, running with unbalanced currents, or from poor connections between the rotor bars and short circuiting end rings. Correct the first two troubles as previously described. The last trouble may be corrected by tightening the connections and brazing the bars to the end ring.

D-C MOTOR TROUBLES

Troubles found only in d-c motors and the suggested procedure to correct the trouble are as follows:

1. Motor attempts to start but overload relay trips. This condition is caused by a weak field or no field. If the motor is an adjustable speed motor, check the field rheostat for correct setting. If correct, check the condition of rheostat. Check the field coils for open circuit and the wiring for loose or broken connections.

2. Motor runs too slow under load. This indicates that the line voltage is too low, that the brushes are set ahead of neutral (if an interpole motor), or that there is an overload. Check the line voltage and correct if necessary. Adjust the brushes to neutral. For overload, check the power input to motor.

3. Motor runs too fast under load. This may be caused by a weak field, too high Une voltage, or brushes set behind neutral (if an interpole motor). If the motor has no inter-poles, speedup is caused by the brushes moving either ahead of or behind the neutral. To check for a weak field, examine the field circuit and rheostat for loose connections; measure the field current. The line voltage should be checked at the motor terminals. If it is too high, correct it. Set the brushes on neutral.

4. Faulty commutation—high, white spark under one brush. Copper embedded in the brush causes this condition. Scrape the copper off with a brush or a knife. Sand "the brush to fit. If necessary, recondition the commutator in one of the manners previously described.

6. Heavy sparking. The many causes for sparking include overload, open field circuit, reversed pole connections, open armature coil, high mica, pitted mica, and brush trouble (including uneven spacing, brushes wedged in holders, brushes worn too short, vibrating brushes, and brushes off neutral). Except for vibrating brushes and high or pitted mica, the method of

correcting the trouble has been discussed or is obvious. High or pitted mica must be undercut. For vibrating brushes, you must check to see if the cause is excessive machine vibration, rough conamutator, incorrect brush pressure, wrong grade of brushes, wrong brush angle, incorrect distance from brush holders to commutator, or high mica.

TESTING ARMATURES

There are two practical tests for locating shorts, open, and grounds in armatures: the growler test and the bar-to-bar test. A growler (fig. 4-7) is a device consisting of a stack of H-shaped laminations cut out on top to hold an armature between its two poles. A coil of wire is wound on the cross bar of the H and is energized with an alternating current. As the current passes through the coil of the growler, a voltage is induced in the armature coils by transformer action.

A growler may be used as follows for locating grounds: Place the armature on the growler and connect one lead of an a-c millivoltmeter to the armature shaft; use the other meter lead as a probe. Turn on the current and touch each commutator bar with this probe. A reading on the meter will be noted for all good coils. No reading on the meter indicates that the coil connected to the commutator bar is grounded.

Figure 4-7.—Growler with armature in position for testing.

To locate an open coil, set the armature on the growler as shown in figure 4-7. Test the top two adjacent commutator bars with an a-c millivoltmeter. If the coil is continuous, there will be a deflection of the meter. Xo deflection indicates an open coil. Keep rotating the armature and testing the two top commutator bars until every coil has been checked.

A shorted coil is located by holding a thin piece of metal, such as a hacksaw blade, directly over the top slot of the armature and along the length of the slot containing the coil side being tested. When the current is turned on, the blade will vibrate rapidly and cause a growling noise if a coil in the slot is shorted.

BAR-TO-BAR TEST

If a growler is not available, you can locate armature troubles by the commutator bar-to-bar test. Troubles are usually confined to one coil or group of coils, and segments to which they are connected can be located by the bar to-bar test. In this method, a low d-c voltage is impressed between two points diametrically opposite on the commutator. The usual equipment for testing smaU machines is a voltmeter or millivoltmeter, a rheostat, and a 6-volt battery. The most accurate readings on the meter can be obtained near the center of the dial; the rheostat is used to control the current flowing through the coils and to keep the meter reading within the desired bounds. Circle the armatiu-e, keeping the test probes on adjacent segments. The magnitude of the reading is not important, since the readings are only for comparative purposes. If the readings are the same, there is no trouble in the coils. A low reading indicates a shorted coil.

To test for an open, adjust the rheostat so that the meter reads on the lower half of the scale; this protects the meter to some extent when an open coil is bridged. If only one coil is open, the readings on one half of the commutator in a 2-pole lap winding are zero except when the meter spans an open coil. When an open is spanned, the meter reading is very high or completely off scale, because the meter completes the circuit.

The test for ground differs slightly from the tests for shorts and opens. Instead of both meter probes being comiected to adjacent commutator segments, one side of the meter is grounded to the armature core and each commutator bar is touched in turn with the other probe. If there is a ground.

there will be two or more bars with practically zero readii^. Some of these grounds are real, and some are phantoms. Mark them all with chalk. Now rotate the armature a few degrees and make another test. Again mark all the bars that indicate grounds. The real grounds will be in the same bars as before, and each bar with a real ground will have two marks, but the phantom grounds shift to other bars which wiU have only one mark each.

TEMPORARY ARMATURE REPAIRS

When a defective coil is removed from an armature, the coil must be replaced with a jimiper of equal current carrying capacity until a new coil is installed. Jumpers can easily be placed on lap windings, because these windings are short and are connected to either adjacent or alternate segments.

On a wave winding, jumper any two of the adjacent segments that indicated trouble when the growler or bar-to- j bar test was performed. A jumper between any two adjacent segments wlll cut out l^o coils in addition to the bad coil. This will lower the efficiency somewhat, but the winding will operate normally for aU practical purposes until a new coil is installed.

This method of jumpering segments can be used on all wave windings that have a brush for each pole. If the winding is multipolar and has only two brushes, good commutation can be obtained by jumpering out only the faulty coil.

REWINDING ARMATURES

Before stripping an armature which may need to be re-woimd, record all available winding data on a card. This information should consist of (1) the nameplate data of the machine, (2) the number of slots, (3) the nimiber of commutator bars, (4) the number of coils in each slot, (6) the number of turns per coil, (6) the size of wire, (7) the coil pitch, (8) the commutator pitch, (9) the end-room measurements, and (10) the type of winding (lap or wave).

After recording the initial winding data, perform a growler

^J• a bar-to-bar test to determine the fault. If a short is indicated, disconnect the top commutator leads and make an insulation test between commutator bars or slip rings. If a short is found, disconnect the bottom commutator leads and test to determine whether the short circuit is in the coil or in the conmiutator. If the short is in the commutator, reinsulate the defective segments; then reconnect the commutator leads and again make a test to determine whether any trouble exists in either the winding or the commutator.

If it is necessary to rewind the armature, first disconnect and remove the coils. During this process accumulate the winding data that was impossible to obtain before stripping. Remove the banding wires by filing them apart. Do this in two separate places. If banding wires are not used, remove the wedges in the slots. A simple means of removing the wedges is to place a hacksaw blade, with the teeth down, on the wedge. Tap the top of the blade to set the teeth in the wedge and then drive out the wedge by tapping the end of the blade. Examine the commutator to determine whether the winding is lap or wave and record this along with the other armature data.

Next, unsolder the coU leads from the commutator, and raise the top sides of the coils for the distance of a coil throw. The bottom side of a coil can now be reached and the other coils can be removed one after the other. Care should be taken to preserve at least one of the coils in its original shape for use as a guide in forming the new coils. The wire size, the number of turns per coil, and the type of insulation on the coils and in the slots should be recorded.

To raise the coils without damaging the insulation, use a small block of wood as a fulcrum resting on the core laminations and a steel bar or piece of wood as a lever. After you have partly raised the coil, drive a tapered fiber wedge between the top and bottom coUs within the slot to finish raising the top coil from the slot. After the coils are removed, clean the slots by scraping and burning the old insulation. File each slot to remove any burrs or slivers and clean the core thoroughly with compressed air.

Before starting the rewinding, you must insulate the core to prevent contact with the coils. Fish paper or fuller board is the usual insulating material. The insulation should extend about Yi inch beyond the slot to prevent the edge of the laminations from injuring the coils. It should be wide enough to fold over the coil after both layers are in the slot. The V-rings (spider) should be well covered with mica or fiber and secured by linen tape. Figure 4-8 shows how an armature core should be insulated.

OR INSULATION PAPER Figure 4-8.—Armature core and insulation.

Small armatures can be hand wound (fig. 4-9). If the armature is too heavy to hold, it may be placed on supports. Larger armatures use a form-woimd coil. Form-wound coils should be wound on a coil-winding machine and pulled into shape on the forming machine. The shape of the coil is determined by the old coil. After the coils have been formed into the desired shape, slip cotton sleeving over the two

Figure 4-9. —Winding armatures by hand.

wires forming the leads, and tape the whole coil with half overlapping tape. The tape may be applied either by hand or by machine. Heat the taped coil in an oven for about 30 minutes, remove, and dip in baking varnish. Now place the coil back in the oven at a temperature of about 220° F,, and bake until the coil is dry. Repeat this treatment as necessary imtil a glossy, varnished surface is obtained.

When inserting preformed coils in the slots, place the lower side first, then the upper side. See that the coil pitch is correct. Insert all the coils until the winding is complete. Place a strip of fuller board in the slot between the upper and the lower coil sides, and a similar strip at the back and the front of the armature where the top and bottom sides cross. If the slot has straight sides, fill it with a strip of hard fiber on top of the coils so that they may be held down by the banding wires. In some armatures the slots are shaped so that fiber wedges may be driven in each slot from the end to hold the coils in place.

Before soldering the coil terminals to the commutator, thoroughly test the winding for grounds, opens, shorts, and reversed and dead coils. Place cotton sleeving over each coil-terminal wire as an added insulation. Take great care duriiig soldering to prevent solder from falling or running down the back of the conmiutator, as this would cause a short circuit. Place the tip of the soldering iron on the commutator near the riser and wait until the iron heats the riser hot enough to melt solder. Touch the solder to the riser and allow the solder to flow down and around the wire and into the wire slot and then remove the iron. Tipping the armature so that the solder will not flow down the back of the commutator helps to prevent short circuits. (See fig. 4-10.)

The ordinary soldering iron does not supply heat fast enough to perform a satisfactory soldering job on a large armatiu'e. You can overcome this difficulty by applying a soft flame from an acetylene torch to the outside end of the commutator bars. The heat travels along the bars to the other end, where you make the connections. Tin the coil ends to be connected to the conmiutator bars with the

Figure 4-10.—Soldering coil leads to commufotor.

soldering iron; tin the bars with heat from the torch and make the connections without applying the flame to the commutator risers. As too much heat can ruin the commutator insulation, wrap the windings in asbestos tape for protection, then heat the armature in an oven for about 2 hours to dry out all moisture. After allowing the armature to cool, dip it in an insulating compound, and then bake it. If equipment for vacuum impregnation of the winding is available, immerse the armature in insulating compound in a closed tank, and pull a vacuum on the tank. This forces the compound around the wires.

To prevent centrifugal force from throwing the coils outward when the armature rotates, wind a band of high-grade steel piano wire on a strip of leatheroid, and place the strip

round the armature and over the coils about 2 inches from lie edge of the core. Do this before dipping or baking the rmature. It is best to perform this operation while the bindings are hot because the insulation shrinks when heated, 3 more flexible, and can be pulled down tiglitly much easier han wlien cold. After the first banding wire is wound on the rmature, insert small tin clips under the wire. Wlien you lave wound the required number of turns on the armature, urn the ends of these clips up over the wires to hold them ightly side by side. Solder these clips with a tin solder, and un a thin coat of solder over the entire band to secure the vires together. In large armatures, notches are provided to iccommodate the banding wire.

Secure the end windings with band wires wound on insu-ating hoods to protect the coils. For the commutator end, place strips of thin mica with overlapping ends on the commutator neck and hold them by a few turns of cord.

A completely rewound armature is shown in figure 4-11.

Figure 4-11.—Rewound armature.

If it is necessary to rebuild a commutator, use molding micanite to insulate between the spider and the commutator. Use commutator mica as insulation between the segments. After assembling the commutator, heat it and tighten with a clamping ring. If shrink rings are provided, do not put them on until the band wires have been tightly placed around the

326605'—55 7

commutator. It is usually more practicable to replace a small commutator rather than to rebuild it.

When the armature is completed, place it in a lathe, and take a very light cut over the surface of the commutator and over the face and side of the commutator tang (riser). This operation ensures that the commutator is perfectly true. The commutator mica should be undercut to a depth of to %i inch so that the carbon brushes will not be damaged by high mica. Undercutting may be done either with a motor-driven tool or a hand tool. A good hand tool for use in undercutting may be made from a piece of hacksaw blade mounted in a handle. Balance the armature on a pair of steel knife edges which are perfectly level and mounted parallel to each other.

The procedure just described applies to practically all d-c armature repair jobs and with slight variations to smaller a-c armatures. In most a-c machines there is no commutator, but extra insulation is required between end turns of adjacent coils when they are connected in different phases. Large a-c motors sometimes have wound rotors, and lai^e a-c generators have wound stators and wound revolving field rotors. The principles of winding are exactly the same, but spare coils are usually furnished already formed and insulated for a-c machines. Such machines usually operate at high voltages, so exercise extreme care when you assemble and insulate these rotors and stators.

RECONDITIONING ARMATURES

Moisture is the principal cause of grounds on armatures and coils. This can come from spray, immersion, or just from operating in a damp space. If an armature or coil has been immersed in salt water, all traces of salt must be removed. The armature or coil must be repeatedly washed in fresh water until no trace of salt can be found on the windings. The armature must then be thoroughly dried by baking and reinsulated if megger tests do not show good insulation between coils and frame. A growler test should also be made for shorted coils. Frequently the insulation of an armature is either oil-soaked, water-soaked, or old and dead, though

none of the windings are bumed out. In such cases the armature should be washed thorou^ly in Stoddard solvent to remove grease or oil, and then dried by heating. Next, it should be treated with baking varnish. Finally, insulation-resistance tests should be made to determine whether all the moisture was expelled from the windings by the heating process.

Armatures should never be baked inmediately after dipping. The varnish must be allowed to drain before applying heat because the varnish forms a skin on the surface and prevents thorough evaporation of the solvent on the inner portion of the varnish film. The purpose of the insulating varnish is to fill the fiber of the insulating fabric for waterproofness, dielectric strength, and flexibility. The finishing varnish suppUes a glossy hard finish that repels oil, moisture, and dirt.

Sometimes it is satisfactory and more economical to replace one or more bumed-out coils rather than to rewind the entire armature. For this purpose prewound coils are frequently furnished in the repair-parts box. Spare form-wound armatures and field coils sometimes are carried as repair parts ready for replacement in lai^e motors and generators.

The conmiutator must be protected against breakdowns. All the joints at each end between the bars and the retaining flanges must be completely filled and sealed with a high-grade baking insulating varnish. In addition, the short creepage paths from the bare copper bars at each end to the bare metal flanges must be protected with a heavy coating of oilproof dielectric paint to prevent shorts or grounds at these points.

REWINDING FIELD COILS

Field coils are divided into two general classes: shunt and series. Shunt coils consist of many turns of fine wire and carry a relatively small current. Series coils are larger coils made up of a comparatively few turns of large-sized wire and carry a relatively large current. In either case the ampere turns of the coil determine the degree of magnetization of the

field poles. Consequently, for coils of comparatively few turns, it is essential that the rewound coil contain the same number of turns as the original coil. The number of turns in a coil of many turns of very fine wire is not particularly important, since a slight error would change the current in the coil only slightly. The shunt field magnetization would not be altered. It is only on series and conunutating windings composed of a few turns of heavy wire that the exact number of turns must be duplicated on the rewound coD because these are in series with the armature and the magnetic effect for a given load cmrent varies directly with the number of turns.

Before unwinding a coil, you should record all the essential data, including (1) the dimensions of the coil, both with the tape on and with the tape removed; (2) the weight of the coil without the tape; (3) the size of wire; and (4) the kind of insulation. If a similar coil in good condition is available, measure its resistance. This value of total resistance combined with the exact wire size provides suflScient data to determine the length of wire in the finished coil. You can accurately determine the proper amount of wire for a shunt coil having many turns of fine wire by obtaining the weight of the copper and the space which the turns occupy. The figures for the weight and the space can be used instead of taking an accurate count of turns.

The equipment for properly rewinding coils includes a lathe or suitable faceplate which can be turned at any desired speed, and an adequate supply of the proper size of wire wound on a spool. A shaft should support the spool to allow free turning, and a weight should be supported on the shaft to provide tension on the wire as it leaves the spool. Secure a core having the exact inside dimensions of the coil to the lathe or faceplate. The core for shaped field pole coils is best made from a block of wood shaped exactly to the required size and provided with flanged ends to hold the wire in place, as shown in figure 4-12. One of the end flanges should be removable so that the finished coil can be taken from the forming block. If the external lead is to be a

92

heavier size of wire than the coil itself, the lead should be spliced to the wire from the spool and thoroughly insulated with treated tape. Short lengths of tape ties, to be used in temporarily binding the finished coil, should be laid in the corner slots of the form.

Next, wind the wire from the spool onto the core or forming block for as many turns as are required. The turns must be evenly spaced one against the other until the winding is completed. In a-c coils, it is well to use insulating paper between layers. Secure the turns by tape, and cut the wire leading to the spool, leaving sufhcient length to make the external connection. Place the inside lead across the coil side, and to provide insulation, insert a piece of fish paper between the lead and the coil. The inner and outer leads should be well insulated from the coil and secured as shown in figure 4-13 to prevent shifting. The terminal leads are usually heavier than the coil ends and are usually made of stranded wue. These leads should have a soldered connection that is mechanically and electrically secure. The coil must be carefully wound with unvarnished cotton or linen tape, and the shape of the coil must be carefully preserved.

Rgure 4-1S.—Forming block for field coils.
Figure 4-13.—Rewound field coil.

When insulated, the finished field coil should withstand twice the normal rated voltage, plus 1,000 volts, to grouiul. This dielectric strength is attained by winding the coil with linen tape, dipping the coil into a baking varnish, and baking for 10 hours at 125° F. Repeating the dipping and baking process improves the insulating qualities of the coil.

If it is necessary to round the field coil to fit the inside curvature of the motor frame, it should be bent into shape before applying the varnish. A more satisfactory method is to make a wooden form having two parts between which the coil is pressed into the required curvature.

Series and commutating pole coils may be wound with ribbon or strap copper instead of round wire. These coils have only a few turns, which are laid in one single turn per layer with mica insulation between layers. The same essential procedure for winding, insulating, dipping, and baking applies to both these coils and shunt coils.

A coil must be tested for polarity, grounds, opens, and shorts after it has been wound, taped, formed, dipped, and baked. Use a small battery to send current through the coil while a compass is held along the axis, several inches away from the coil. If the south pole of the compass needle points toward the middle of the coil, the coil face nearest tlie

compass is a north pole. This coil should be placed on a north pole in the same position as it was during the time of the compass test, and the field current should flow through the coil in the same direction.

The polarity test automatically indicates whether or not the coil has an open. To test for a short, current should be sent through all the field coils in series and the voltage drop should be measured across each coil separately. The new coil should have about the same voltage drop as the other coils. If the voltage drop is less than the other coils, the new coil has a short, or too few turns in it. If the voltage drop is greater than the average of the other coils, the new coil has too many turns or the wrong size of wire has been used. A growler may also be used for a quick check for shorts.

REWINDING STATORS

On an a-c motor, tests may indicate that the stator winding is defective and needs rewinding. For ordinary repair an exact reproduction of the winding is necessary; therefore, it is extremely important to keep an accurate record of all pertinent data concerning the winding If possible, obtain the data before stripping the stator. If this is impossible, obtain the information during the stripping operation. The necessary information consists of (1) the nameplate data of the machine, (2) number of poles, (3) number of slots, (4) coil pitch (the number of slots each coil spans), (5) number of turns in each coil, (6) size of the wire in each winding (starting and running), (7) type of connection, (8) determination of the position of each winding in relation to the other (if there are two windings), (9) type of winding (hand, form, or skein), and (10) slot insulation.

STRIPPING THE STATOR

You can strip an a-c armature or stator in much the same way as a d-c armature. You must save one coil in order to provide the dimensions for the new coils. Be sure to measure the end room of the coils before they are removed from the slots. Record the distance and take care that the new coils do not extend farther than this distance from the ends of the slots. As most stator windings are hard-baked, you will usually find it necessary to cut the winding on one side of the stator and pull the wires through the other side. If you apply a blowtorch to the winding or if you heat it in an oven you can soften the insulating material and facilitate stripping.

WINDING THE COILS

For small motors the coils ma^^ be wound by hand into stator slots, but most coils are prewound and preformed. The size and shape of the coils are best determined from the coil that was removed intact. With this coil as a model, use a winding board or form to duplicate, exactly, the shape of the windings. For small motors you may wind coils into rectangular forms and then stretch them into a diamond shape by pulling at the centers of opposite sides. When you have wound a coil, tie it in several places to hold the turns together. Then remove it from the form and tape it with cotton or varnished cambric tape.

PLACING STATOR COILS IN THE SLOTS

When all the coils have been wound, taped, and formed, you are ready to place them in the stator. The slots must be insulated with the same size and type of insulation that was used in the original winding. In an open-slot stator, the coil sides are placed in the slot intact, but in a semienclosed slot the turns of the coil side are spread out and the coil is held at an angle so that all the turns may be fed into the slot (fig. 4-14). Make certain that each turn is placed inside the insulation. Sometimes wires are placed between the insulation and the iron core by mistake so that a groimd results. Pull the side of the coil through the slot until all the turns are in the slot. This one coil side occupies the bottom half of the slot. The free end coming from this side is known as the bottom lead of the coil. The other coil side remains

Figure 4-14.—Placing a coil in a semienclosed slot

free, as shown in figure 4-15. Continue by placing one side of the second coil in the slot beyond the fii-st slot. The remaining coils are fitted in the same manner until the slots spanned by a complete coil pitch each holds one side of a coil. The second side of each coil is left out of the slot until the bottom half of the slot is occupied by a coil side. The second side of a coil is then fitted on top of the first side of a coil several slots awaj*, depending on the pitch of the coil.

Therefore, one side of each coil is in the bottom half of a slot, and the other side of that coil is in the top half of another slot several slots awaj*. The number of coils of which the top side is left out is usually one or two more than the coil pitch. These coils should not be put into the slots until all the bottom sides are installed. Make certain that each coil side extends beyond the slot at both ends and does not press against the iron core at the corners. Before inserting the second coil side in a slot, it is necessary to insulate it from the coil side already in the slot. This can be done by placing a strip of insulating paper, slightly wider than the slot, over the bottom coil. Extra insulation must be placed between the pole-phase groups because of the high voltage existing between phases. The computations indicated in the section headed "stubbing" should be completed before

INSULATION PLACED ON TOP OF SLOT TO PROTECT WIRE FROM SCRAPING IRON CORE

, SLOT I

I r SLOT 2

BOTTOM SlOe OF COIL
FIRST COIL OF WINDING IN PLACE
TOP SIDE NOT IN SLOT

1st COIL L. ONE SIDE
Z.4 COIL —1^ ^'•^^T SECOND COIL OF WINDING IN PLACE Figuft 4^15.—Coil sldtt placed In tloh.

any coils are inserted in the stator so that plans for this extra insulation can be made.

With all the stator coils in place, force down the coils in the slot securely by placing a piece of soft wood or fiber on top of the coils through the slot opening and gently tapping on the windings. Place another strip of insulating paper over the coil windings and then insert fiber wedges to hold the coils in place.

Three-phase motors usually have as many coils as there are slots. These coils are so connected that they produce three separate windings or phases, each of which must have the same number of coils. The number of coils in each phase is found by dividing the total number of coils in the motor by the number of phases. Also, all three-phase motors have their phases arranged in either a wye or a delta connection.

A wye-connected three-phase motor is one in which the beginning of each phase is connected to the Uve or external circuit and the ends of each phase are connected together. A delta connection is one in which the end of each phase is connected to the beginning of the next phase. Thus, the end of phase A is connected to the beginning of phase B] the end of phase B is connected to the beginning of phase C; and the end pf phase C is connected to the beginning of phase A. At each of these three points a connection is made to the external circuit.

The niunber of coils in each pole is determined by dividing the total number of coils by the number of poles. A i>ole-phase group is a definite number of adjacent coils connected in series. In all three-phase motors there are always three groups in each pole, one for each phase, such as one group for phase A and one group each for phases B and C. Therefore, if a pole has nine coils there must be three coils in each group. Such a section consisting of three coils is known as a POLE-PHASE GROUP. The coils of any group are always connected in series. Thus, the end of coil 1 is connected to the beginning of coil 2, the end of coil 2 is connected to the beginning of coil 3, the beginning of coil 1 and the end of coil 3 are coil-group leads for connection to other groups. \Vhen the number of coils in each group is known, the coils can be connected into groups. The number of pole-phase groups is determined by multiplying the number of poles by the number of phases.

STUBBING

When all the coils have been inserted in the stator slots, the ends insulated, and the slot wedges driven into place, there will be two free ends for each coil (top and bottom) extending from one side of the stator winding. You must connect these free ends so as to form a series of groups of coils. The number of coils to be connected into any one group depends on the total number of coils used, the number of phases in the supply circuit, and the speed of the machine. The number of coils in each group is equal to the total num-

ber of coils in the stator divided by the number of pole-phase groups. For example, in a machine that has 36 stator slots and 36 coils, the winding is to be arranged for three phases and 4 p>oles. If these 4 poles are to be formed in each phase, there are necessarily 3X4, or 12, pole-phase groups. As there are 12 pole-phase groups, it follows that there are 36-r-12, or 3, coils in series in each group. Therefore, you must connect three consecutive coils together in series for each pole-phase group by (1) twisting the adjacent ends together, (2) soldering them, and (3) taping them to form a stub between coils.

In connecting the three coils, connect the outside end or lead of one coil to the inside end or lead of the next coU, the outside end of that coil to the inside end of the next, and so forth. Use the inside end of the first coil and the outside end of the last coil in connecting one pole-phase group to another pole-phase group. In connecting the coils into groups, start at one place and bend the inside lead of the first coil in toward the center, and then bend the outside lead of that coil and the inside lead of the next coil together. Connect the outside lead of the last coil with the inside lead of the next one, and bend away the outside lead of this coil from the center. Repeat this process for each of the pole-phase groups all around the stator. If you should make an error in dividing the coils into phase groups, you would discover it and could easih' correct it before connecting the coils together.

Next remove the surplus insulation, such as the cotton covering and sleeving, from those wires that are to be twisted together in connecting one coil to the next. If the copper wire is tarnished or enameled, you must polish it with sandpaper and tin it. Repeat this process for all the leads to be used in connecting the coils into phase groups. Grip the ends of the two leads with pliers and twist them together. The length of the twisted portion should be about an inch. After twisting the ends together, check the winding to see that the proper number of coils have been connected together. Now coat the twisted ends with a soldering flux and solder them. Cut off the ends of the wires so that the soldered ends

100

or stubs are about Yi- to 1-inch long. Insulate the stubs by slipping flexible varnished cambric tubing or sleeving over the stubs. Cut off the tubing about % inch beyond the ends of the stubs. If the distance to the bearing brackets or frame of the machine is small, bend the insulated stubs inward between the coils so that they do not come in contact with the frame when the stator is assembled in the machine.

The next step is to connect the different pole-phase groups together. You can connect the groups of coils of one phase so they are all in series, are all in parallel, or are in a combination of series and parallel groups, depending entirely on the number of pole-phase groups in each machine.

CONNEaiNG POLE-PHASE GROUPS

Before connecting the pole-phase groups together, construct a diagram containing the pole-phase groups in each phase and the number of poles for the particular motor, as shown in figure 4-16. To be certain that the pole-phase

Figure 4-16.—^Thrce-phase four>pole winding for a 36-slot stator.

groups are connected properly to obtain alternate north and south poles, indicate the direction of current flow througb each pole-phase group by arrows, which must reverse in direction for each successive group. Now you can connect the pole-phase groups into phase groups.

TESTING THE STATOR

After the stator has been rewound and the connections have been made, you should test it for grounds, opens, shorts, and reversed connections. Test the stator before it is varnished. Trouble may be corrected more easily before varnishing.

Grounds may be detected by use of a megger. If a ground is detected, it must be removed. You may need to disconnect each pole group and localize the ground to one group. Then break down the grounded-pole group to the individual coils. After you have located the grounded coil you may bt-able to reinsulate the coil in the slot. If not, you must remove it.

Opens are located by using an ohmmeter and checking the coQ ends. Poor stub connections between coils are frequently the cause of open stators. Any coil that does not indicate a reading is open and must be rewound. An open circuit in the starting winding may also involve the centrifugal switch, so you must check this thoroughly to ensure (1) clean, bright contacts, and (2) sufficient pressure on the contacts.

Rough handling of coils while they are being inserted into slots may result in damage to the insulation between turns of the same coil and thus may produce a short circuit. Shorted coils are located most easily in a rewound armature by using an internal growler. This growler is the same as the one previously described except that it is shaped to fit the inside of a stator, as shown in figure 4-17. The growler is placed on the core of the stator and is moved from slot to slot. A shorted coil produces a very heavy current, which is indicated by the rapid vibration of a hacksaw blade held

at the other end of the coil. If a growler is not available, another way in which to locate the shorted coil is to pass a low value of direct current through the winding. Then touch a screwdriver blade to the surface of the stator and move it around the inner circumference. You can detect the shorted coil by the weaker magnetic pull it has on the screwdriver blade.

Figure 4-17.—^Testing with internal srowler.

Reverses result from incorrect connections between poles and are best discovered by means of a polarity test. Apply a low d-c voltage across the two ends of one of the three-phase windings and hold a compass inside the stator; move the compass slowly from one pole to another. The compass needle reverses itself at each pole if the stator is correctly connected. If the same end of the compass needle is attracted to two adjacent poles, a reversed pole is indicated.

VARNISHING AND BAKING

•When aU the connections between coils, groups, and pole-phase groups have been completed and tested, and when the flexible leads to the power supply have been attached, the stator or armature is ready for varnishing.

The varnish for dipping and baking should be at a tem- ' perature of 77° F. to 90° F. If the repairs have been made under conditions of high humidity, give the equipment a prehminary baking, but allow it to cool in the oven before dipping. Inmicrse it in the baking varnish until bubbling ceases. The first dip usually requires 15 minutes, and each additional dip requires about 5 minutes. Rotate the equip-ihent slowly until aU parts are well covered and bubbling ceases. Then remove it from the varnish and allow it to drain until all dripping stops. At this time, wipe the met«l parts such as the commutator and slip rings with a cloth dipped in thinner. If your shop is equipped with a vacuum i impregnation tank, apply a partial vacuum while the coils | are immersed in the varnish. This draws out entrapped air and permits the varnish to penetrate the winding when the vacuum is released.

After the dripping stops, place the equipment in an oven at a temperature of 266° to 275° F. Bake for about 6 hours for each dip except the last, which should be for about 15 hours. Two dips usually are suflBcient. The time of the bake specified is general and may have to be modified according to the size of the particular apparatus.

GROUNDING ELEaRICAL EQUIPMENT AND APPLIANCES

One very important job you must now include in your repair procedures, and in addition to your inspection and installation operations, is the effective grounding of equipment and appliances at your base to protect personnel against shock. For some time past, standards have required the grounding of enclosures of equipment and appliances operating from supply circuits in excess of 150 volts such as X-ray machines, heavy equipment machines, and transformer and generator station equipment. Only in special cases, however, was grounding recommended when the supply circuit was less than 150 volts. SecNav Instruction 5100.2 of 25 September 1953 has extended these precautions to cover lower voltages. It directs that each activity of the

Shore Establishment inaugurate a planned program of grounding the enclosures of all applicable electrically operated equipment and appliances powered from circuits in excess of 50 volts and less than 150 volts a-c or d-c. Grounding in the range less than 50 volts is left to the discretion of local authority and it is up to you to find out what the policy is on your station. The instruction also directs that the ground connections be mechanically secure and electrically adequate, and that the equipment and appliances be checked periodically at intervals not in excess of one year to ensure that insulation has not deteriorated to an unsafe condition. BuDocks feels that an insulation resistance of 50,000 ohms is intended to be a realistic value; however, for portable hand tools, efforts should be made to maintain a minimum resistance value as near 100,000 ohms as practicable. The SecNav Instruction further requires that equipment and appliances found to be unsafe be taken out of service until repaired.

In accordance with the instruction the following appliances and pieces of equipment must have their cases or enclosures grounded:

Hand held appliances, such as: Office Machines, such as —Con.
Drills Calculators
Floor polishers Dictating machines
Floor scrubbers Mimeograph machines
Grinders Paper cutters
Industrial-type vacuum Typewriters
cleaners Sanders Saws
other similar tools Refrigerators
Sewing machmes
Office machines, stich as: Stoves
Adding machines Washing machines
Addressograph machines
Miscellane<yus:
Record players Extension lights, when attached
Radios to portable cords
Television sets Radio noise interference sup>-
Clocks pressors
Desk lamps, etc., when enclosed Soldering irons
in metallic conducting surfaces Solder pots
Heavy appliances, such as: Ironers
326605'—55 8
Mitcellantom —Continued
Glue pots, etc. Clothes cutting machines Hair clippers Hair drj ers Immersion heaters
Permanent wave machines Portable ventilating fans Radiant heaters
Vibrating machines (massaging)
Galley equipment and appliances, such as:
Clothes dryers Coffee grinders CofFeemakers Food choppers Food freezers
Broilers
Food mixers Grills Hot plates Toasters
Vending machines Waffle irons
Other similar appliances

This list is not intended to be all-inclusive. Your station will have to survey its own additional equipment to determine whether or not it must be grounded.

Installation standards, grounding methods, technical requirements, and other information required to fulfill the provisions of the establishing instruction may be found in BuDocks Instruction 5100.3 of 28 January 1955. In order to comply with this instruction you must be on the lookout for any supplementary instructions and notices in this series which may be issued from time to time in reference to this safety program. It is your responsibility to see that all exposed noncurrent carrying metal parts such as metallic conduit, cable sheath or armor, appliance enclosures, switch boxes, and the like, on applicable equipment and appliances be kept as near ground potential as possible, by providing a low resistance to ground. Do not ground to steam or fire sprinkler systems; use such grounds as water pipes, rigid conduits, and ground rods.

1. What determines the layout of an electrical repair shop?

2. Who has authority to order changes in shop layout diagramed in Advanced Base Drawing?

3. State the goal of any inspection schedule.

4. State two general rules for successful motor and generator main-

QUIZ

tenance.

5. At what point on the commutator should the brushes of a direct current machine be set to obtain the best operation?

6. Name the two abrasives which may be used in seating brushes.

7. Why should carbon tetrachloride not be used for cleaning of the commutator?

8. Give two reasons why a light cut should be taken when a commutator is turned down in a lathe.

9. List five motor troubles common to both a-c and d-c motors.

10. Name two tests for locating shorts, opens, and grounds in armatures.

11. What step should be taken before an armature that may need to be rewound is stripped?

12. When coil terminals are being soldered to the commutator, how do , you prevent the solder from flowing down the back of the commutator?

13. If it is necessary to rebuild a commutator, what material should be used to insulate between the spider and the commutator?

14. What causes most grounds on armatures and coils?

15. What is the purpose of insulating varnish on an armature?

16. What are the two general classes of field coils?

17. What data should be recorded before a field coil is unwound?

18. How is dielectric strength for a field coil attained?

19. After a field coil has been wound, taped, formed, dipped, and baked, what tests must be performed?

20. After a stator has been rewound and connections have been made, what tests should be made?

CHAPTER 5

INSTALLING COMMUNICATIONS EQUIPMENT

COMMUNICATIONS AT ADVANCED BASES

Communication facilities at an advanced base are far less elaborate than facilities at permanent bases located in the United States. The equipment must be rugged and as compact and portable as possible.

The types of communication equipment commonly used at advanced bases include telephone systems, interoflSce communication equipment, and public address systems. Ad ability to follow wiring diagrams and a thorough knowledge of the principles of electricity will enable you to install practically any type of electrical equipment. The Seabees have demonstrated this fact over and over. An example is the 600-line telephone exchange installed at the Naval Air Facility, Port Lyautey, French Morocco. Six CE's without previous training for such a job installed this complex system within 8 weeks.

Even though such tasks can sometimes be performed without specialized training of personnel, it is desirable that everyone understand equipment with which he will be working. This chapter discusses communication equipment that is used at advanced bases.

TELEPHONE SYSTEMS

Telephone systems are classified according to the method of switching used at the central ofiice. The two basic systems are the dial system and the manual system. In the dial system, connections between telephones are set up automati-

cally by electromechanical switching apparatus at the central office. These connections are controlled by the operation of the dial on the calling telephone and are made without the aid of an operator. A dial central office requires much intricate wiring and complex equipment and is rarely used at advanced bases.

In the manual telephone system, the connections between telephones are made manually by an operator at the switchboard. Manual telephone systems are further classified according to the sources of electrical energy supplied to the transmitters. The two basic methods of supplying this energy are the local-battery system and the common-battery system. Both types are used at advanced bases.

LOCAL BAHERY SYSTEM

In a local-battery system, the sources of electrical energy for the transmitters and signaling devices are included in the telephone sets located at each telephone station. The term LOCAL indicates that the sources of electrical energy for the transmitters and for signaling are a part of each individual telephone station. A local dry-cell battery supplies the current for the transmitter circuit, and a hand generator (often called a magneto) supplies the current for signaling.

Local-battery telephones may be directly connected to each other without going through a switchboard. In such a case, a transmitter and receiver, a local battery, a ringer, and a hand generator are necessary for each telephone. Usually, local-battery connections between telephones are made through a switchboard.

In addition to a generator, battery, transmitter, and receiver, the switchboard is equipped with jacks, plugs, and drop signals. The drop signal is an electromechanical device with a shutter which falls and attracts the attention of the operator when the user of a telephone signals with his hand generator.

Local-battery switchboards used by the Seabees generally have either 12- or 40-line capacity.

COMMON-BAHERY SYSTEM

In a common-battery system, all the stations and the switchboard obtain the necessary electrical energy for transmission and signaling from a source of power at the central office. This source of power is a power line, a generator, or a large storage battery. A ringing machine attached to the power source supplies ringing current for all the telephone sets. Automatic signal lamps, rather than drop signals, are the devices used to signal the operator when a receiver is removed from a hookswitch.

Common-battery systems may be interconnected with local-battery systems, thus permitting telephone conversations between a station connected to a local-battery switchboard and one connected to a common-battery system. Some switchboards operate on a combination of common-battery and local-battery power. The 57-line board used by the Seabees is an example. The other two types in general use at advanced bases are a 12-Iine board and a 40-line board, both local battery types. The 12-line and 57-line switchboards are discussed in succeeding paragraphs.

12-LINE SWITCHBOARD

The 12-line switchboard used by the Seabees is a portable, local-battery type. It is designed to establish a working telephone system for units up to and including the size of a battalion. The model in general use is the SB-22/PT (fig. 5-1). This set is also used by the Army Signal Corps.

The SB-22/PT 'is a small, lightweight, immersionproof switchboard used for interconnecting local-battery lines. It requires no special mounting equipment for operation. The unit consists of a battery case, an operator's handset-headset, an accessory kit, and a switchboard assembly. The assembly consists of a metal case containing 36 binding posts, 12 line jack telephone circuits (commonly called line packs), and one operator's telephone circuit (commonly called operator's pack). A removable front cover provides access to the control and operating portions of the plug-in units. A

Fisure 5-1.—12-line swUchboard^ SB-22/PT, front view.

rear access door provides access to the 36 binding posts located on the frame. (See fig. 5-2.) These binding posts are used to make connections to the s\vitchl)oard.

GROUND BATTERY IC£NT FICi'^C'N
BINDING ="'ST
Fisore 5-2.—12-line switchboard, SB-22/PT, rear view.

The front cover is secured to the case by two spring-locking devices, one located on each side of the case. A rubber gasket attached to the front cover provides a waterproof seal when the cover is latched. A circuit label is located on the inside of the front cover and a writing surface for making notations of calls in progress is provided on the outside of the cover. The rear access door is hinged at one end and is secured by recessed thumb latches at the other end. A writing surface is provided on tlie rear door for identification of the binding posts. The case has two openings, one on each side, to provide an entrance for the wire. A slotted rubber gasket covers the opening and prevents the entrance of moisture, dust, and other foreign matter.

The SB-22/PT switchboard is also equipped for remote control radio communication. Under normal conditions you will be concerned only with the telephone lines. The procedure for connecting the wires of the switchboard to a radio circuit is very similar to the procedure for making connections to the telephone circuits.

Lin* Pgdct

The 12 line packs are located on the left side of the front panel. A line pack is shown in figure 5-3. Each line pack is fastened to the switchboard by two captive screws, one at the top of the unit, the other at the bottom of the unit. The line pack consists of a reel unit, a drop, a jack, and an identification strip.

The REEL UNIT consists of a reel, a cord, and a plug. The cord is fastened to the spring-loaded reel by three screws at one end and is equipped with a standard switchboard plug at the other end. The cord may be extended to a maximum distance of 35 inches and is retracted by the spring-loaded reel.

The DROP consists of an electromagnet and a hemispherical piece of metal with a luminous strip painted horizontally across it. When a telephone user turns the crank of the hand generator serving the telephone, a circuit is completed,

SNAP-ON COVER
1-201 Switchboard signal PS01 Cord assembly
J201 Telephone jack ES01 Reel unit
CR201 Selenium rectiRer E202 Lightning arrestor
P20S Clip assembly Rgure 5-3.—Line pack for SB-22/PT switchboard.

causing the hemispherical piece to rotate downward; tliis exposes the himinous strip. The dl'op Is restoied by a m^ ohanical linkage botwoon the drop and the jack

The JACK is used in conjunction with the cord of the operator's pack, or witli cords of the hne packs, to interconnect hue circuits.

Fisure 5-4.—Operator's pack for SB-22/PT switchboard.
114

The IDENTIFICATION STRIP is a piece of white plastic fastened to the front of the line jack. Marks on this strip identify the telephone circuit associated with the line pack.

The operator's pack (fig. 5-4) is located on the right of the front panel. It is fastened by four captive screws, one at each comer. The pack consists of a reel unit, ringing equipment, alarm equipment, and a handset-headset receptacle.

The REEL UNIT is identical to that of the line pack. The ringing equipment consists of a hand generator and a control switch. Ringing current is generated when ths hand crank is rotated. The control switch has two positions: ring back and PWR RING FWD. This switch enables the operator to connect ringing current to the caUing or to the called telephone.

The ALARM EQUIPMENT providcs a more positive signal for the operator than that afforded by the drop mechanism of the line pack. The alarm equipment consists of the signaling devices and a selector switch. The signaling devices are a buzzer and a lamp. When the switchboard is signaled by an outlying telephone, the drop on the hne pack associated with the outlying field telephone falls; this completes a circuit through the alarm lamp or the buzzer dependingon the setting of the switch. Either the lamp lights or the buzzer sounds.

The HANDSET-HEADSET rcccptacle is a polarized, bayonet-locking, 10-conductor receptacle which is used with the plug on the handset-headset cord.

Operartor's Pack
NOMENCLATURE—Fisurc 5-4.
P302 P303
1-301 Nijht alarm buzxtr T301 Induction coll G301 Rlngins 9«ntrator
P305 P306i
P304> Cllpancmbly
S30S Nisht alarm twitch
S303 Lever switch (ringing)
J 301 Receptacle connector
1-302 Night lamp
S301 Signal lamp twitch

C301 Coupling capacitor

L301 Retardation coil

E301 Reel unit

C302 Coupling capacitor

Prcinstallation Procedures and Checks

The 12-line switchboard may be placed in any convenient location. It should be placed on an operating surface of the proper height to ensure ease of operation. If a suitable stand is not immediately available, it may temporarily be placed on the ground.

Be careful not to damage the equipment during uncrating and unpacking. Inspect it inmiediately upon unpacking for possible damage during shipment. Save the original packing cases and containers. They can be used again when the equipment is repacked for storage or shipment.

Before making the connections, use the following procedures to test the line packs:

1. Place the tip and sleeve of the operator's plug across the binding posts associated with the line pack being tested.

2. Ring on the line with the hand generator; the drop should operate. If it does not operate, check all wiring and repeat the test.

3. Insert the plug of the operator's pack into the jack of the line pack under test. The drop should restore. If it does not restore, replace the line pack.

4. Extend the cord of the line pack under test about one foot and release it. It should retract completely into the jack of the line pack. If it fails to retract properly, replace the line pack.

6. Repeat these tests for each of the 12 hne packs.

Test the operator's pack as follows:

1. Connect a field telephone which is known to be in operating condition to any line pack which has tested as good.

2. Insert the plug of the operator's pack into the jack of the line pack.

3. Rotate the switchboard hand generator crank with the

RING BACK-PWR RING FWD SWitch in the RING BACK pOSitioD.

The ringer of the field telephone should operate. If it fails to operate, check all the wiring and repeat the test.

If, after the second test, the riiiger still does not operate, replace the operator's pack.

4.. Remove the plug of the operator's pack from the line pack. Operate the alarm switch for the visible alarm condition and operate the hand generator crunk on the telephone. The lamp should light. If it does not light, replace the lamp.

5. Repeat step number 4 as a test for audible alarm condition.

6. Insert the plug of the operator's pack into the jack of the line pack and remove the handset from the hook-switch of the telephone connected to the line pack.

7. Direct someone to talk over the line and note the quality of transmission. If transmission is poor or speech cannot be heard at normal speech level, check the wiring and repeat the procedure. If the trouble continues, replace the operator's pack.

Installation Procedures

Make the groimd connection before connecting the wire lines to the switchboard. Select the lowest, dampest site in the vicinity; clay or loamy soil is best. Scoop out a hole 6 inches deep and drive a ground rod into the hole until the top of the rod is about 3 inches below the undisturbed ground surface. Connect one end of a wire to the ground rod. Saturate the earth around the rod with water, and fill in the hole with earth. Connect the other end of the wire to the binding post located on the frame behind the rear access door.

Use the following steps to connect the wire lines to the switchboard:

1. Select a pair of binding posts for connection to the desired circuit.
2. Record the circuit identification on the proper identification strip.
3. Separate the wires of the twisted pairs connected to the outlying telephone.
4. Strip the insulation ⅜ inch from the end of each conductor.
5. Place a screwdriver into the slot at the top of the binding post.
6. Depress the binding post with the screwdriver.
7. Insert the bared end of the wire into the hole at the top of the binding post.
8. Release the top of the binding post. The wire should be fastened firmly in the binding post. Apply tension on the wire to test the connection.
9. Repeat the same procedure for the other wires. When the wire lines are connected, you are ready to

install the 4 dry-cell batteries, which furnish the power. Remove the battery case by pulling it out of the spring-loaded cUps. Unscrew one end and insert two BA-30 batteries into the compartment so that the brass contact end of each battery faces outward. Repeat the procediu-e for the compartment at the other end of the battery case. After completing this operation, replace the case into the spring retaining clips.

Stackins of Two Switchbooidt

To serve more than 12 but fewer than 30 lines, you may stack two 12-line switchboards. Remove the operator's pack from one board, and install five line packs in the empty space. Place this modified switchboard on top of a normally equipped switchboard. It is necessary to use two jumpers between the two switchboards. Connect one jumper between the binding posts marked NA and another between the posts marked GND. Pass the jumpers through the slot at the side of each switchboard. One set of batteries serves both boards; remove the battery case from the one containing the 17 line packs. The field telephones can then be connected as previously described. A maximum of 29 lines may be served.

Maintenance and Repair

Frequent inspections are necessary to keep the system in good operating order. Dust, dirt, grease, or moisture, should not be allowed on the exterior of the switchboard or the handset-headset. Use a clean, dry, lint-free cloth or a dry brush for cleaning them. Clean the electrical contacts on the frame of the switchboard with a cloth moistened with dry-cleaning solvent; wipe them dry with a dry cloth. Use a burnishing tool to clean the switch contacts.

Rust, fungus, dirt, and moisture tend to accumulate on the binding posts, plugs, and external portions of the line jacks. Remove with a dry rag or brush. Inspect the battery case and clip contacts for corrosion, moisture, fungus, and tightness. Check all lines for kinks, strains, moisture, fungus, and loose terminals; see that the insulation is not frayed, cut, or damaged.

Moving parts of the SB-22/PT do not require lubrication. When necessary, lubricate the spring latches of the front cover, and the thumb latches and hinge of the rear access door.

Failure of the switchboard to operate properly is usually caused by one or more of the following faults:

1. Rundown batteries
2. Defective operator's handset-headset
3. Defective line pack
4. Defective operator's pack

Rundown batteries may cause any of these diflSculties:

1. Pull switch is pulled out, but lamp fails to hght.

2. Drop of a line pack indicates that a telephone is signaling the switchboard, but the lamp or buzzer of the operator's alarm fails to work.

3. The operator cannot talk to any of the telephone users. Remove rundown batteries, and replace them in the

manner previously described.

A defective handset-headset prevents the operator from talking to any of the telephone users, from receiving calls

from any of the outlying phones, or both. Replace a defective handset-headset with one that is in working order.

A defective line pack may be responsible for any of the following troubles:

1. The user of a telephone cannot signal the operator because the drop does not function.

2. The drop of a line pack for a particular telephone works, but the operator's alarm fails to function. (If this difficulty occurs with all telephones, the trouble may be rundown batteries or a defective operator's pack.)

3. The operator cannot talk to or receive calls from the user of a telephone.

4. The users of two telephones cannot converse with each other, but the operator can converse with each user separately.

The line pack is a self-contained, plug-in unit. It may be removed by loosening the two captive screws from the front panel, inserting the plug into the jack, and pulling on the plug until the line pack is free from the front panel. The replacement should then be inserted into the empty space. After the captive screws are tightened, the new line pack is ready for operation.

A defective operator's pack may cause any of the following difficulties:

1. Lamp fails to light when a pull switch is pulled out (also may be caused b\'7d- burned-out lamp or rundown batteries).

2. The operator's alarm fails to work when the drop of a Une pack indicates that a telephone is signaling the switchboard (may also be caused by bumed-out batteries).

3. The operator cannot ring any field telephone.

4. The operator cannot talk to any telephone users (may also be caused by a defective handset-headset or bumed-out batteries).

5. The operator cannot receive calls from telephones on the line (may also be caused by a defective handset-headset).

6. The operator cannot ring back.

To replace an operator's pack, loosen the four captive screws in the front panel; then, grasp the ringing generator crank and pull out the unit. Insert the replacement pack into the front panel, and tighten the captive screws.

57.LINE SWITCHBOARD

The 57-line switchboard used by the Seabees is a part of telephone central office set TC-2, an Army Signal Corps-type equipment. This switchboard consists of 20 magneto

DIAL AND MANUAL TRUNK ANSWERING JACKS

COMMON BATTERY LINE JACK AND LAMP ASSEMBLY

DIAL.THROUGH LINES, AND CONFERCNCE JACKS

MAGNETO LINE JACK AND SIGNAL ASSEMBLY

OPERATORS JACKS

SWITCHBOARD TEST AND OUTGOING Line JACKS

Figure 5-5.—Switchboard BD-89-C. 326005"—55 -» ^^l

(local-battery) lines and 37 common-battery lines. A switchboard of this type is designed for a base where the majority of telephones are in a centrally located area, but where a few telephones are located in outlying activities. Telephones located a considerable distance from the exchange require local-battery operation.

The complete switching system consists of the 57-line switchboard, a main distributing frame (MDF), a pwwer panel, a rectifier, a power unit, and accessories. Advanced Base Drawing No. 529,373 shows a layout for this equipment.

The switchboard is the Army Signal Corps-type BD-89. It is a portable, single-position, two-panel, manually operated type. In addition to the 57 telephone lines, it contains three trunkllies which may be connected to other switchboards. One trunk circuit provides for two-way service between a dial central office. A dial cord circuit is provided for dialing on the trunk circuit. Lamp signals are provided for the common battery lines and the cord circuits.

Drop signals are provided for the magneto lines, with magneto recall lamps associated with each cord circuit. The line jacks are wired to a terminal strip at the rear of the switchboard which provides for connections with spade-terminal strips through rubber-jacketed cables to the main distributing frame. Terminal facilities are provided for making connections to the storage battery supply, ringing current supply, and grouping key circuit. One model of the BD-89 switchboard is shown in figure 5-5.

Main Distributing Frame

The main distributing frame serves to terminate the outside lines and to connect them to the proper line jacks on the switchboard. Cabinet BE-79 (fig. 5-6) is the MDF used with the BD-89 switchboard. The MDF is equipped with protector blocks and heat coils which are connected to terminal strips. Binding posts are provided for the incoming lines and for cross-connecting, so that all such connections

can be made without soldering. The connections from the MDF to the switchboard are made with rubber-jacketed cables equipped with binding post cable connectors (terminal strips) at the cabinet, and spade-terminal cable connectors (terminal strips) at the switchboard. Twelve repeating coils are mounted on the line side of the MDF to permit cross-connection and cabling.

Figure 5-6.—Cabinet BE-79, main distributing frame. A. Line side. B. Switchboard side.

The line side of the cabinet contains two vertical strips of protectors, each strip consisting of 40 pail's of protector blocks and heat coils. The heat coils guard against the cumulative effects of small currents that might produce excessive heat when they flow for considerable periods of lime. A terminal panel equipped with 80 binding posts for connecting the incoming lines is mounted on the line side of the protectoi-s. The central office side of the protectors is permanently wired to a terminal panel equipped with 80 binding posts from which the lines may be cross-connected to the switchboard cable connector binding post terminals or repeat coils.

The switchboard side has three cable connectors. Each connector consists of a strip of insulating material equipped

with a row of 50 binding post terminals (25 pairs) to which the switchboard cable conductors are soldered. These soldered connections are enclosed in a metal protecting cover. The switchboard cable consists of three rubber-jacketed cables, each containing 25 pairs of braid-covered, latex-insulated, conductors. The cables, exclusive of cable connectors, are 21 feet long. Each cable connector at the switchboard end of the cable is made up of two strips of insulating material between which is mounted a row of 50 spade terminals. The spade terminals are so mounted that some movement is allowed; this enables the spade terminals to be self-alining when connections are made to the binding posts on the switchboard terminal panel. The cable conductors are soldered to spade terminals. The soldered connections are enclosed in a metal protecting cover.

The 12 coils on the coil rack are mounted directly below the protectors in two horizontal rows of six coils per row. Each coil is mounted separately to simplify removal if*a replacement is required.

Other Auxiliary Equipment

Other auxiliary equipment necessary to operate the switchboard includes a power service panel, power unit, junction box, accessories (headset, chestset, ground rods, and cords), rectifier, storage batteries, and a control panel. The latter^ serves both to control the rate of chaise of storage batteries and to provide ringing current to the switchboard.

The power service panel (fig. 5-7) is designated cabinet BE>-75. In figure 5-7, it is mounted on a rack (designated FM-30) along with a rectifier and filter reactance. Cabinet BE-75 contains two sockets and two outlets for the connections of the rectifier and the other power cords. Three circuit breakers control the current to the sockets.

The power unit will most likely be a gasoline engine driven a-c generating set designed to generate 120-volt, single-phase, 60-cycle current. The power unit is used only if another source of power is not readily available.

Figure 5-7.—Power service panel, cabinet BE-75.

The junction box (JB-19) provides a means of bringing in 24 volts direct current, 20-cycle, a-c ringing power, to the switchboard. It may also be used to extend power to a second switchboard and to group the current between the two switchboards. The junction box should be mounted inside the switchboard on the upper right hand side. Connections should be made to the terminal box through three rubber-covered cables which terminate in a cable connector on one end, and on the switchboard terminal strip on the other end.

The headset and chestset make up the operator's telephone set and are used by the operator for talking and listening during operation of the switchboard. The ground rod is used to terminate connection to ground required for protection of the switchboard equipment. The cords provide the connections between the various pieces of equipment such as power and power panel, ground rod and MDF, and batten and control panel.

The rectifier serves to convert alternating current into direct current, which is necessary to operate the switchboard, and also for charging the storage batteries, which serve as an emergency source of power. The rectifier will most likely be of the selenium-disk type, but the vacuum-tube type is also used.

The panel which controls the rate of charge of the storage batteries is designated panel BD-98 (fig. 5-8). The control panel also contains two interrupters which are used to supply ringing current to the switchboard. Interrupter PE-25fl

CONNECTOR FOR INTERRUPTER PE-250 OUTPUT TO SWITOMBOABD
CONNECTOR FOB INTERRUPTER PE-24e OUTPUT TO SWITCHBOARD
Figure 5-8.—Control panel, BD-98.

supplies the current when a-c power is available, and interrupter PE-248 provides ringing current when only battery power is available.

Installation ProcedurM

The location of the switchboard controls the location of the other equipment in the telephone system. The switchboard should be placed in a location that will give the operator maximum signal visibility. It should usually be located with the face at right angles to the window to prevent direct light from striking the eyes of the operator. A space of at least 30 inches should be allowed between the back of the switchboard and any wall or obstruction; this permits easy access to the rear for maintenance and repair. A clearance of at least 2 feet should be maintained at a side of the switchboard that is to be used as a passageway. A minimum of 40 inches should be allowed between the keyshelf and the nearest wall or obstruction; about 6 feet is considered ideal.

After you have set the switchboard in the desired location, use the following steps to complete the installation:

1. Loosen the two screws holding the bottom panel on the front of the switchboard and remove the panel.

2. Remove the cloth securing the cord weights and restore the panel.

3. Remove the front cover.

4. Loosen the mounting screws at the extreme ends of the drop retaining strips and raise the retainer strips from the locking position to the service position, leaving the drops free to operate.

5. Loosen the two screws on the back panel and remove the panel.

6. Insert the resistance lamp in the socket in the upper left side of the switchboard.

7. Insert the adapter in one of the two outlets on the bottom of the switchboard and install a heating element in the adapter.

8. Install a second heater element in the socket mounted . on the lower portion of the equipment panel. The socket is

mounted so that the heating element will be parallel to the bottom of the switchboard.

9. Mount the junction box, JB-19, on the upper right side of the cabinet. Two holes in the bottom of the junction box fit two dowel pins on the framework of the switchboard. Tighten the two screws which fit into the top of the junction box.

Set up the auxiliary equipment as follows:

1. Mount the control panel (panel BD-98) on the base of case CS-71, the container in which the panel was packed. Secure the panel to the case with six bolts, nuts, and washers which are shipped with the component.

2. Remove the ground cord from its packing and connect it to the ground terminal on the panel frame. The ground terminal is located on the back of the panel in the upper left corner.

3. Remove the four screws holding interrupter PE-250 to the panel. Tilt the interrupter far enough forward to permit the insertion of the lamp in the socket on the inside of the interrupter. Return the interrupter to its original position and restore the four screws.

4. Set the top portion of case CS-71 in its desired location with the open side up. Place panel BD-98 on top of the case; , the bottom of the panel (base of case CS-71) will fit into and I be held in place by the eight comer reinforcing boards. i

5. Mount rack FM-30 (shown in fig. 5-7) on top of control : panel BD-98; secure with four mounting bolts.

6. Mount the rectifier on rack FM-30 and secure with mounting bolts.

7. Mount the filter reactance on the rear of rack FM-30; secure the reactance to the middle and the bottom cross bars.

8. Mount the power service panel, cabinet BE-75, on the rear of rack FM-30; secure to the top and middle cross bars.

After setting the MDF in the proper location, remove the two covers; they are held by four screws, one in each corner. Remove the bag containing repair parts, the spare cord, and the silica gel from the cabinet. Disconnect the switchboard end of cord CD-298 from the cabinet.

Set the two storage batteries on rack FM-31 and place the rack within cable distance of control panel BD-98. Place the power unit within cable distance of panel BD-98.

If possible, the ground source should be a water pipe or similar buried metallic object of good conductivity, having a large area of earth contact. If such a ground source is not available, drive three ground rods (GP-29 or MX-148/G) deeply into moist earth at least 10 feet apart. Use light strokes to prevent whipping of the rod from destroying the earth contact. Tamp the soil firmly around the rods. Moistening the surrounding earth with salt water improves the ground conductivitj'.

Mqkins the Connections

The connections between components of the set are shown in figure 5-9. If 110-volt, 60-cycle, a-c power is available, connect cord CD-393 to the source and to the connector in the center of the top panel of cabinet BE-75. Tighten the connector until it fits snugly, but do not force the connector and ruin the threads. If the power described is not available, connect cord CD-409 between the connector top of power panel BE-75 and the outlet box on the power unit.

Connect the two storage batteries together, using cord CO-38. Connect the batteries to the connector on control panel BD-98 marked bat cable, using cord CD-335.

Connect the front cord of the rectifier to the connector on panel BD-98 marked BAT CHG. Now connect the back cord to 1 of the 2 top connectors on the power service panel,* cabinet BE-75.

Connect the cord from interrupter PE-250 to 1 of the 2 top connectors of cabinet BE-75.

Connect the ground cord of panel BD-98 to the ground terminal screw on the MDF, cabinet BE-79. Use cord CC)-258 to connect the ground terminal of cabinet BE-79 to the ground rod.

Moimt the junction box JB-19 in the switchboard and fasten the switchboard spade terminal cable connector to the junction box. Connect the smaller of the two junction box
BE-75TO

Figure 5-9.—Cording diagram of telephone central office let TC-2.

cables to the connector on panel BD-98 marked teller. Connect the larger cable to the connector on the panel marked SWBD BAT CA.

Connect the outlet box (mounted inside the switchboard near the middle of the right side) to 1 of the top 2 comiectors on either side of cabinet BE-75; use cord CD-395 to make tlic connection.

Preventive Maintenance

A preventive maintenance program may be divided into four basic operations: (1) Inspecting, (2) tightening, (3) cleaning, and (4) adjusting. Inspection is the most important of these operations. Minor defects, which do not interfere with the performance of equipment, but which may lead to a major breakdown, can usually be located during a careful inspection. Check for the following conditions:

1. Overheating as indicated by discoloration, blistering, or bulging of the parts or surface of the container; leakage of insulation compounds; or oxidation of metal contact surfaces.

2. Placement, by observing that all leads and cable are in their original positions.

3. Cleanliness, by examining all recesses in the units for . accumulation of dust, especially connecting terminals. Parts, connections, and joints should be free of dust, corrosion, and other foreign matter. In tropical and high humidity locations, look for fungus growth and mildew.

4. Tightness, by testing any connection or mounting that appears to be loose.

Daily Inspection

Make a daily inspection of the switchboard exterior, night alarm bell, headset, chestset, exterior of MDF, storage batteries, and control panel BD-98. Wipe off any dirt or dust with a soft, dry cloth; never use soap and water. Clean the surface of the switchboard key shelf with a soft bristle brush. Pull the switchboard cords up as far as possible and let them hang down over the key shelf. Dust along the cord rail, being careful to prevent foreign objects from getting into the cord sockets.

The night alarm bell should be inspected whenever a new operator comes on duty. Test by operating the NA key to the ON position and allowing one of the drop shutters to fall to the operated position. The night alarm bell will sound if the system is working properly.

Examine the exterior of the headset and chestset for dirt, dust, rust, and corrosion. In addition, check for chipped

paint or fungus growth on the chestset. In wiping off the chestset, be careful to keep dirt or lint from getting through the holes into the face of the transmitter.

In examining the MDF, check for damaged places, chipped paint, dirt, dust, rust, corrosion, and loose, or missing screws. When tightening the screws, be careful not to force them.

Make the same checks on control panel BD-98 as on the MDF; in addition, inspect the cable connections to the panel, and examine interrupter PEl-248 and the meters for cracked or broken glass. Clean dirt, dust, and Unt from the wiring on the rear of the panel with a soft bristle brush.

When inspecting the batteries, check the specific gravity and level of the electrolyte in addition to the check for cleanliness and tightness of connections. Remove corrosion from the battery terminals with fine sandpaper or crocus cloth.

Weekly Inspection

Weekly inspections should be performed on the drop shutters, switchboard cords, keys, fuses, protector blocks, heat coils, repeating coils, and ground rods. Examine the drop shutters for bent or damaged latches or bent hinge pins. If necessary, adjust the shutter latches with the long-nose pliers to prevent the shutters from falling to their operating position when the switchboard is jarred. Bend the latch so that the shutter wiU fall freely when a call is received, but will not drop to the operating position when jarred.

The switchboard cords should be examined for dirt, dust, mildew, and fungus. Check the cord weights and pulleys for smooth operation.

Inspect the keys for tightness of mounting and loose, cracked, or broken handles. In tightening the two screws that hold the keys in place, be certain to use the proper size screwdriver. A screwdriver that is too large damages the screw slots and the bakehte cover. Tighten the key handles with the fingers, being careful not to exert enough pressure to cause the handles to crack.

Check the fuses for correct capacity and tightness of mounting; tighten mounting screws securely, but be careful not to damage the fuse by using too much force.

Examine the protector blocks for cracked or broken porcelain and carbon blocks, dirt, dust, and foreign matter. Replace any blocks that are chipped or broken. Clean the blocks with a soft bristle brush; remove any foreign object which may be lodged between the protector blocks. Use extreme care when brushing or you may dislodge a block, thus causing trouble on an incoming line.

Inspect the heat coils for cleanliness and for chipped or broken shells. Clean with a soft bristle brush. Replace any that are defective.

The repeating coils should be checked for cleanliness and corrosion. Do not remove the metal covers of the coils. Examine the coil mountings for loose, damaged, or missing screws. Clean the coils with a clean, dry cloth.

Inspect the ground rods for rust and corrosion. See that the wing bolt at the terminal connection is tight.

Monthly Inspection

Plugs, relays, capacitors, terminals, binding posts, and cables, should be given a monthly inspection. Clean the switchboard plugs with cord plug polish, using a clean, dry cloth. Remove all residue of the polish after cleaning in order to maintain good electrical contact.

The relays and capacitors need to be inspected only for dirt and foreign matter. Clean them with a dry cloth.

Inspect the terminals aad binding posts on the switchboard and control panel for cleanhness and tightness. Check the incoming Hne connections for good electrical contact. Tighten any loose terminals with a suitable screwdriver or wrench. Clean the terminals and binding posts with a soft bristle brush.

Examine connecting cables for damaged or worn insulation. Check the fittings on the ends of the cables for tightness and good electrical connection. Tighten the connections on the cables as required.

Troubleshooting and Repair

When trouble in a telephone system arises, the first stop in effecting repair is to determine the probable cause of trouble. Schematic diagrams are useful in localizing the fault to a particular component. Make a complete visual inspection of the wiring and connections to the associated equipment. If no wires or connections are broken, the trouble must be located by makii^ continuity, voltage, and resistance measurements. By following the circuit with a systematic process of elimination, you can usually locate the fault within a short time. Start at the point where the analysis has shown the circuit to be good and proceed step by step, eliminating parts of the circuit, until the fault is located.

If visual inspection fails to reveal the source of trouble, you will have to use electrical testing equipment. Study the schematic and wiring diagrams, and make tests until the trouble is located.

The troubleshooting chart shown on succeeding pages will simplify fault location. It is designed for locating trouble in models A, B, C, D, E, and G of the BD-89 switchboard. Various troubles are listed, together with the. probable location and the recommended correction. By using tliis chart, you can frequently isolate the trouble to one part of the equipment, thus saving time-consuming checks on trouble-free components.

You can effect many repairs simply by replacing a fuse, lamp, cord plug, strap, or similar item. Other repair* involve nothing more than cleaning a contact, cord plug, or grouping key. Adjusting a screw, a spring, or line drop will often make equipment serviceable.

Some repairs, of course, involve rather complex operations; examples are repairs to switchboard cords and repairs to combined jacks and signals. Space does not permit a discussion of all such repairs. The Army Technical Manual, TM 11-340, which describes the TC-2 telephone set, contains a complete discussion on all phases of maintenance and repair for the switchboard and associated parts.

Troubleshooting Chart for 57-line Switchboard

Symptoms

No transmission or reception on 1 line circuit.

2. No transmission or reception on any line circuits.

3. No transmission on any line circuit.

Probable troubles

4. No reception on any line circuit.

5. Operator cannot be signaled on 1 common battery line circuit.

Open in associated switchboard wiring.

Dirty or defective

jack contacts. Open in cord or cord

plug.

Short in cord or cord plug-

Talk-ring key not making proper contact.

Open induction coil in operator's circuit.

Operator's jack dirty

or defective. Open in operator's

headset, cord, or

plug-Operator's jack dirty

or defective. Open in operator's

headset, cord, or

plug.

Open or shorted

transmitter. Operator's circuit

fuse blown. Open in operator's

headset, cord, or

plug.

Open or shorted receiver.

Lamp burned out

Dirty or open contact in jack.

Open in associated switchboard wiring.

Corrections

Repair open.

Clean, repair, or replace jack.

Repair or replace cord or cord plug.

Repair or replace cord or cord plug.

Clean, repair, or replace key.

Replace induction coil.

Clean, repair, or replace jack.

Repair headset.

Repair or replace cord or plug.

Clean, repair, or replace jack.
Repair headset.
Repair or replace cord or plug.
Repair or replace I transmitter.
Replace fuse.
Repair headset.
Repair or replace cord or plug.
Repair or replace receiver.
Replace lamp.
Clean contact.
Repair or replace jack.
Repair open.
Troubleshooting Chart for 57-Line Switchboard—Con.
Symptoms
Probable troubles
Corrections
6. Operator cannot
be signaled on any common battery line.
7. Operator cannot
be signaled on 1 magneto line.
8. Permanent signal in cdmmon battery line.
9. Operator cannot signal on 1 line circuit.
10. Operator cannot signal on any line circuit.
Blown fuse
Main circuit breaker open.
Open or short in line drop coil.
Open or short in associated switchboard wiring.
Dirty or open contacts of line jack.
Line drop out of adjustment.
Short in associated switchboard wiring.
Carbon protector blocks on ring side of line grounded. Open, ground, or short in associated switchboard wiring.
Dirty or defective jack contacts.
Open or short in cord or cord plug.
Dirty cord plug
Talk-ring key not making proper contact in ring position.
Open in Talk-ring key wiring.
Interrupter PE-250 or interrupter PE-248 not operating properly.
Dirty or open contacts of RINGING key.
Replace fuse. Restore circuit breaker.
Replace combined
jack and signal. Repair open or short.
Clean or adjust contacts. Repair drop.
Remove short.
Replace defective protector blocks.
Repair open. Remove short ground.

or

Clean, repair, or replace jack.

Repair or replace cord or cord plug.

Clean plug.

Clean, repair, or replace key.

Repair open

Check and repair interrupter PE-250 or interrupter PE-248.

Clean, repair, or replace key.

Troubleshooting Chart for 57-Line Switchboard—Con.

Symptoms

11. Operator cannot signal on trunk to manual common battery exchange.

12. Cannot signal operator on trunk from manual common battery exchange.

13. Operator cannot signal on trunk to dial exchange.

Probable troubles

Dirty contacts or defective line jack.

Dirty contact or defective dial jack (Switchboard BD-89-G).

Open in associated switchboard wiring.

Dirty contacts or defective A relay (Switchboards BD-89-A, -B, -C, -D, and -E).

Dirty contact or defective dial jack (Switchboard BD-89-G).

Line drop defective or out of adjustment (Switchboard (BD-89-G).

Dirty contacts or defective C relay (Switchboards BD-89-A, -B, -C, -D, and -E).

Burned-out answering lamp (Switchboards BD-89-A, -B, -C, -D, and -E).

Open in associated switchboard wiring

Dial jack contacts dirty or not making properly.

Open or short in dial cord or cord plug.

Dirty or defective contacts of dial.

Corrections

Clean, repair, or replace jack.

Clean, repair, or replace jack.

Repair open.

Clean, repair, or replace A relay.

Clean, repair, or replace jack.

Adjust or replace drop.

Clean, repair, or replace C relay.

Replace lamp.

Repair open.

Clean, repair, or replace jack.

Repair or replace dial cord or cord plug.

Clean, repair, or replace dial.

326605»—55 10

Troubleshooting Chart for 57-Line Switchboard—Con.

Symptoms

14. Cannot signal operator on trunk from dial exchange.

15. No night alarm on

1 magneto line circuit.

16. No night alarm on

. any magneto line circuit.
17. No night alarm on
any common battery line circuit.
18. No night alarm on
any line circuit.
19. No supervision on common battery line circuits (1 answering cord).
Probable troubles
Dirty contacts or defective dial jack.
Line drop out of adjustment or defective (Switchboard BD-89-G).
Answering lamp burned out (Switchboards BD-89-A, -B,-C,-D,and-E).
Dirty contacts or defective C or D relay (Switchboards BD-89-A, -C, -D, and -E).
Open or short in associated switchboard wiring.
Contact at drop dirty or not making.
Dirty or open contact of NA key.
Dirty or open contact of NA key.
Dirty or open contacts on relay A or B.
NA fuse blown
Dirty or open contacts on night alarm bell.
Wmding of night alarm bell open or shorted.
Answering supervisory lamp burned out.
Dirty contacts or defective A or BS relay.
Corrections
Clean, repair, or replace dial jack. Adjust or repair drop.
Replace lamp.
Clean contacts. Repair or replace C or D relay.
Repair open or remove short circuit.
Clean contact. Adjust night alarm springs.
Clean contacts. Repair or replace key.
Clean contacts. Repair or replace key.
Clean contacts. Repair or replace relay A or B.
Replace fuse.
Clean contacts. Adjust contact springs.
Repair or replace bell.
Replace lamp.
Clean contacts. Repair or replace relay.
Troubleshooting Chart for 57-Line Switchboard—Con.
Symptoms
20. No supervisor! on common battery line circuits (1 calling cord).
21. No transmission or
supervision on 3 consecutive cord circuits.
22. No recall signal on
magneto line circuits (I cord circuit).
23. One conference jack
inoperative.
24. Two or more con-
ference jacks inoperative.

25. Through line cir-
cuit inoperative.

26. Outgoing line circuit inoperative.

Probable troubles

Calling supervisory lamp burned out.

Dirty contacts of TALK-RING key.

Dirty contacts or defective C or FS relay.

Blown cord circuit fuse.

Recall lamp burned out.

Open connection to lamp.

Dirty or defective jack contacts.

Short or open in cord or cord plug.

Dirty cord plug

Open in common strap between jacks.

Open in common strap between jacks.

Dirty or open contacts in either jack.

Open strap between binding posts on terminal panel.

Dirty or defective jack contacts.

Short or open in cord or cord plug.

Dirty cord plug

Short or open in associated switchboard wiring.

Corrections

Replace lamp. Clean contacts.

Clean contacts. Repair or replace relay.

Replace fuse.

Replace lamp. Repair open.

Clean, repair, or re-replace jack.

Repair or replace cord or cord plug.

Clean plug.

Repair or replace strap.

Repair or replace strap.

Clean, repair, or replace jack.

Repair or replace strap.

Clean, repair, or replace jack.

Repair or replace cord or cord plug.

Clean plug.

Repair open or remove short.

Troubleshooting Chart for 57-Lme Switchboard—Con.

Symptoms

Probable troubles

Corrections

27. Operator receives interference, crosstalk or noise.

28. Operator's monitoring circuit inoperative.

29. Grouping circuit inoperative.

Grounded line or cord circuit.

Dirty or grounded carbon protector blocks.

Dirty contacts or defective monitoring key.
Open or shorted winding of monitoring coil.
Grouping key not making proper contact.
Open or short in associated switchboard wiring.
Remove ground.
Replace defective protector blocks.
Clean contacts. Repair or replace key.
Replace coil.
Clean, repair, or replace key.
Repair open or remove short.

OTHER SWITCHBOARDS

In the past, the Seabees have used other sizes of switchboards, such as the 6-Hne and the 50-Une sizes. Some of these boards are still in stock. It is expected that other models will be developed in the future.

The procedures for the installation of different models vary somewhat, but the basic principles are the same. The general rules that apply to one local-battery set can be expected to apply to all local-battery sets. The same is true of common-battery systems. Therefore, it is safe to say, that if you can install the two switchboards just described, you can install any switchboard of comparable size.

INTEROFFICE COMMUNICATION SYSTEMS

An interoffice communicating system is used to transmit orders and information among offices that are only a short distance apart. Frequently such offices are in the same building. Intercoms are not used at all advanced bases; if they are used, and there is no Electronics Technician attached to the base, the job of installing, maintaining, and repairing them usually falls to the CE.

Assembling an intercom set requires considerable knowledge of the principles of electronics. For this reason, intercom sets intended for use at advanced bases arc packed ready for operation. By observing a few simple rules and following the wiring diagram that accompanies the set, you should be able to make the installation without difRculty.

An intercommunicating sj^stem consists of one or more master stations, a junction box, one or more remote speaker units, and the wire necessary to make the connections. One type of intercom set used by the Seabees is shown in figure 5—10. The master station has a capacity of 12 remote speaker-microphone units; however, if one or more of the remote units are master stations, the capacity of the system is only 11 remote stations.

The basic parts of the master station consist of a 3-tube chassis, a speaker-microphone, and a selector switch panel. Tlie parts are installed in a wooden cabinet. A combination

Figure 5-10.—Intercommunication set. A. Master station. B. Remote unit.

volume control and ON-OFF switch is mounted directly below the selector switch panel. The pilot light is illuminated at all times when the switch is on. A 3-positioQ switch at the center of the cabinet front controls talk-list«D or idle position. The speaker microphone is moimted inside the cabinet behind the grill on the front panel. A junction box, used for interconnection to remote stations is attached to the chassis by flexible cable. An a-c power cord is attached to the chassis.

The switch panel consists of the selector switches (12 for the model shown), a space above each switch for identify ing the station, and an annunciator for each switch. (Ail intercoms are not equipped with annunciators.) The switches have three positions: OFF, ON, and a third position to operate annunciatoi's on a remote annunciator master station unit.

The annunciators are solenoid plungers mounted above each station selector switch. When the button atop a remote speaker microphone is depressed, or when a switch on a remote n^ster unit is pressed down, a buzzer at the master station sounds, and the annunciator above the switch for the calling station springs outward. The annunciator remains out until the call is answered; it should be pushed back to its normal position at the same time that the selector key is raised to answer the call.

The talk-listen lever is a 3-position switch. The three positions are idle, listen, and talk. Under normal operating conditions, the lever should be left in the idle position. The idle position should be used to determine whether a station to be called is in communication with another station. Leaving the lever in the idle position, flip the selector switch of the station that is to be called. If no other station is conversing with this station, press the lever to the talking; position and speak into the master station. If the system has only one master unit, you may press the lever into the talk position without going through the idle position, since remote speaker microphone units cannot communicate with one another.

Installation Procedures

Any combination of master stations and speaker microphone units up to the capacity of the master station can be used. Where it is not necessary for remote stations to communicate among themselves, only one master station will be installed; this is the usual case.

Install the master station within reach of a 110-125 volt, 50-60 cycle a-c power outlet. The master station and the speaker microphone units should be placed on the desks or in the working spaces of the personnel who will use them. If any of the units are installed outdoors, take the necessary precautions to protect them from adverse weather conditions.

The size of the wires to be used in making connections between units is governed by the length of wire. For the voice lines, use No. 22 to No. 19 twisted pair wire. The maximum wire resistance permissible will be stated in the operating instructions. For the model shown in figure 5-10, the maximum resistance is 50 ohms per pair. The amount of wire determines the wire size to be used. No. 22 wire give a resistance of 32 ohms per 1,000 feet, and No. 19 wire gives a resistance of 16 ohms per 1,000 feet. Use the larger sizes of wire (lower number) when great amounts are necessary to connect the units.

The wire resistance of the annunciator lines must be kept below 15 ohms per pair. You will normally use No. 14 wire (which gives 4 ohms per 1,000 feet) or No. 16 wire (which gives 8 ohms per 1,000 feet).

Make all connections to a master station unit on the junction box. Solder the wires to their respective terminals on the terminal strips. Make connections to the speaker microphone units by removing the back of the cabinet and attaching the connecting wires to terminal screws on the unit. On models having annunciators, attach the annunciator wires to the pushbutton terminal block.

Be sure that the interstation wires do not cross hot pipes. Never place the wires in a location where they are in danger of being covered by water.

After the wiring is installed, check the resistance with an ohmmete'r. Make certain that the maximum permissible resistance is not exceeded and that there are no opens, grounds, or shorts.

Maintenance and Repair

Many of the maintenance and repair instructions that apply to SAnitchboards apply equally to intercom systems. In general, preventive maintenance techniques consist of five steps:

1. Feel
2. Inspect
3. Tighten
4. Clean
5. Adjust

The feel operation- is necessary to check and determine if electrical connections or bushings are overheated. Feeling indicates defects requiring corrections.

Inspection is, of course, the most important opeiig^ion in the preventive maintenance program. Check for the same four conditions that you check on a switchboard:

1. Overheating
2. Placement
3. Cleanliness
4. Tightness

The tightening, cleaning, and adjusting operations are self-explanatory.

The chart shown lists the most common types of troubles that you are likely to encounter in an intercom set.

Components in intercom sets are readily accessible and may be easily replaced if found to be faulty. When a defective component, such as a burned-out resistor or transformer, is located, remember that the cause of the condition may be in some other part of the circuit. If the cause is not located and corrected, the new part will be burned out in the same way as the one that was replaced.

Troubleshooting Chart for Intercom System

Symptoms

Probable troubles

Corrections

1. Pilot lamp does not indicate and tubes do not light when switch is on. Unit does not operate.

Unit operates in talk-listen position of talk-listen lever but no calls can be received from other stations when in idle position.

' 3. Low volume and distortion with talk-listen switch in talk or listen position. 4. Unit operates properly but there is 60-cycle hum in listen position.

Power supply (110-125 volts, 50- to 60-cycle) not on.

Switch on volume control defective.

Poor connections between power plug and pK)wer transformer.

Selector switch or talk-listen switch contacts.

Intersection wires open or shorted.

Defective tube.

Open lO-mf capacitor (C-4) across bias resistor (R-3).

Defective tube

Open or leaky filter capacitor (C-6 or C-7).

If 60-cycle hum results when a selector key is operated, there is a ground or leak in unit or in external wiring.

Test.

Short out to test.

Check circuit with ohmmeter.

Check circuit with ohmtneter. Ohm-meter should read about 35 ohms through speaker if switch (S-2) is in 50-ohm position or 30 ohms if in 500-ohm position. Poor contacts Nos. 4 and 13 on talk-listen switch.

Replace with tube known to be good.

Replace filter capacitor (C-4).

Change tubes. Replace.

Check from wires to chassis or ground with ohmmeter. Resistance should be over 100,000 ohms.

Troubleshootii^ Chart for Intercom System—Continued

Probable troubles , Cwrections

I

SymptoQU

5. Pilot light operates and tubes light but unit does not operate in talk or listen position.

6. Unit operates properly on some stations but not on others.

7. Feedback howl or hum.

8. Distortion and low volume.

9. Dbtortion

Defective tubes

B supply shorted or open.

Open or shorted resistors.

Open contacts on talk-listen switch.

Check wiring circuits to these units with ohmmeter for opens and shorts.

Input and output wiring too close. Capacitor (C-7) open.

Capacitor (C-3) open or leaky; tube (V-1 or V-2) gassy.

Low B supply voltage.

Check and replace.

Check circuit for opens with ohm-meter.

Check all voltages.

Check by shorting each pair of contacts with junnper.

Check contacts on selector switches. Clean with dry-cleaning solvent.

Keep these circuits isolated. Replace capacitor (C-7).

Replace capacitor (C-3) or tube (V-1 or V-2).

Check "B" supply with ohmmeter.

Before unsoldering a part, note the position of the leads. If the part has several connections to it, tag each of the leads. Be careful not to damage other leads by pulling or pushing them out of the way.

It is important to remember the foUow^ing facts when you solder connections:

1. A carelessly soldered connection may create a new fault.

2. A poorly soldered joint is a very difficult fault to locate.

3. It is easy to allow drops of solder to fall iato the set.

4. Drops of solder in a set may cause a short circuit. These points are particularly important to remember

when you are soldering the leads on intercoms. However, this information applies to soldering of all commimications equipment.

PUBLIC ADDRESS SYSTEMS

During the early stages of an invasion, portable types of public address systems are used to amplify speech in the landing area. Small types are d-c battery powered and are completely self-contained. When great sound coverage over a high level of noise is required, a larger a-c portable type, powered by a gasoline driven generator is used.

At an established base, a public address system may be used for an auditorium, outdoor movies, or for camp communications. During World War II, a talk-back type of system was established for camp communications. Horns serving as loudspeakers were placed at strategic locations around the base. A called station could answer the master station if the speaker stood within about 30 feet of the loudspeaker.

The talk-back type is seldom used now. The system generally used for advanced base communications is a portable set consisting of a 100-watt cabinet-type amplifier, a dynamic (movable coil) microphone with heavy-duty floor stand, two 25-foot lengths of shielded microphone cable, and one 25-foot length of heavy-duty power cable. This system requires 100/125 volts a-c on 50/60 cycles.

The horns serving as loud speakers can be controlled individually or in any combination. The speaker can address only one station, a few stations, or all stations. A changeover switch is provided to allow signal input from either a microphone or a phonograph.

As with any electrical circuit, the wiring diagram provides the key for the proper wiring connections. Normally, No. 14 size wire should be used for wiring connections. The horn loudspeakers may be mounted on top of buildings, on poles, on speaker stands, or even in trees. Before making the location of the loudspeakers permanent, it is desirable to test for uniform loudness, for minimum echo, and for dead spots. Follow the recommendations of the manufacturer closely when you make the installation.

Trouble in a p-a system is frequently nothing more than a

loose cable connection or a break in the cable shield. Before commencing lengthy tests, check for faults of this type. In soldering connections make sure that both metals are clean; the completed soldering job should be firm and durable. Faulty soldering can cause faults in the system that are very difficult to locate.

Serious troubles in the system require signal tracing equipment such as an audio-signal generator and an output meter or an oscilloscope. In testing the electric circuit, the most important point to remember is that the trouble should be locaUzed and isolated. A careful study of the circuit diagram will save much unnecessary testing.

SPECIAL SCHOOLS

Each type of communication system discussed in this chapter is a field of study in itself. Principles of sound engineering and electronics, as well as principles of electrical circuits are involved. Because of the specialized knowledge required for communications equipment, the Seabees frequently send CE's who may work in the communications field to special schools. One such school is the Army Signal Corps Laboratory at Fort Monmouth, N. J.

Even after instruction at a school, you will find that you have a lot to learn. There is no substitute for experience.

QUIZ

1. What are the two basic types of telephone systems?
2. How is the source of energy supplied for a local-battery telephone?
3. Describe a drop signal.
4. Where do the stations in a common-battery system obtain the electrical energy for

transmission and signaling?

5. What devices are used to signal the operator in a common-battery set?

6. What is the line jack telephone circuit in the 12-line switchboard commonly called?

7. What is the operator's telephone circuit in the 12-line switchboard commonly called?

8. How many lines can be served when two 12-line switchboards are stacked?

9. On the 12-line switchboard, what is the probable cause when a drop fails to function?

10. What type of switchboard is designed for a base where the majority of telephones are in a centrally located area, but where a few arc in outlying areas?

11. What is the function of a main distributing frame?

12. Name the main auxiliary equipment for the 57-line switchboard.

13. Name the piece of equipment whose location controls the location of other telephone equipment.

14. How can the conductivity of the earth surrounding the ground rods be improved?

16. What are the four basic operations of a preventive maintenance program for a switchboard?

16. What portions of the 67-line switchboard system should be inspected daily?

17. What is the first step in effecting repair in a telephone system?

18. What are the basic parts of an intercom system?

19. Name the most important operation in a preventive maintenance program.

20. For what purposes are public address systems used at an established advanced base?

21. What is the most important point to remember in searching for faults in a public address system?

CHAPTER 6

TELEPHONE CABLE SPLICING TELEPHONE CABLE

Both telephone cable and power cable often require splicing. In this chapter telephone cable sphcing is discussed. Chapter 10 takes up power cable splicing.

Both telephone cable and power cable are usually lead sheathed. The size of the telephone cable indicates the number of pairs of conductors in the cable. The size of a power cable indicates its power-carrying capacity.

Generally the wires in telephone cable are insulated with paper, an effective insulator. During the process of manufacturing cable, the paper is wrapped spirally or sprayed on the wires as pulp. Rubber composition is often used for insulating telephone cable to be used for special purposes.

Paper-wrapped wires in cable may be either double-wrapped or single-wrapped. As a rule they are color coded. For example, in a double-wrapped pair the inner insulation on the wire might be red and the outer layer white. The mate wire of this pair would be inner layer white with outer layer red.

Double-wrapped paper insulation is sometimes called high dielectric insulation. The additional thickness of paper and the added air space between the wires and between the wires and the sheath give extra protection against a breakdown of the insulation caused by excessive current in the cable; for example, an open-wire telephone line coming in contact with a power line and carrying current into the cable.

TAPE-ARMORED CABLE

Ordinary lead-covered cable for aerial or underground construction is known as outside cable. The lead sheath protects the insulated wires inside the cable.

Tape-armored cable consists of lead-covered cable to which a protective covering of paper, jute, and steel tape is added. This type of cable is intended for installation in the ground without conduit, when the increased shielding provided by the steel tape against inductive interference is desired. In addition, steel tape provides some measure of protection against mechanical injury, such as disturbance by heavy construction equipment. Steel tape-covered cable may also be used to minimize damage by rodents. It may be obtained to match practically all standard lead-covered cables.

Tape-armored cable is sheathed with lead-antimony alloy. The protection over the sheath consists of a covering of paper, a bedding of jute, one layer of steel tape, and an outer covering of jute. The paper and jute are impregnated with a preservative compound. As each layer is added to the cable it is completely covered with an asphaltic compound. The finished cable is given a coating of whitewash to prevent the cables from sticking together.

OTHER CLASSIFICATIONS

Lead-covered telephone cable may be further classified as quadded cable, color-coded cable, and composite cable.

A quadded cable is one in which all or some of the conductors are arranged in quads or groups of two pairs to provide suitable characteristics for the use of phantom circuits. (A phantom circuit provides the equivalent of an additional telephone channel by certain arrangements of the existing circuit.)

The term color-coded cable, when used to describe outside cables, indicates that the larger cables (100 pairs or more) are color coded according to groups to simplify splicing. Color coding is used in the description of inside and switchboard cable to indicate that the insulation is colored in accordance with a standard color scheme so that each pair and each unpaired single wire can be identified.

In the description of lead-sheathed cables, composite cable indicates one in which two or more gages of conductors are included.

PLASTICCOVERED TELEPHONE CABLES

Plastic-covered telephone cables are a comparatively recent development. Lead is a relatively scarce and expensive material, so it is not improbable that the use of plastic-covered cable will become more widespread. One of the main advantages of plastic-covered cable is its light weight, which makes it especially suitable for installation at advanced bases.

Splicing techniques and procedures are rapidly being improved and developed for plastic-covered cable. In general, procedm-es for splicing this type of cable closely parallel those used to splice lead-covered cable. The various manu- [f acturers of plastic-covered cable usually issue splicinginstruc- i tions. You should follow these instructions as they will incorporate the latest information.

BUILDING AND SWITCHBOARD CABLE

Building cable is used to furnish wiring for a larger number of circuits than can conveniently or economically be supplied by twisted pairs. Paper-insulated conductors carry the service into the building. A silk and cotton cable is spliced onto the terminating end to prevent moisture being absorbed by the paper-insulated cable.

Switchboard cable is used to connect the switchboard apparatus to the terminal room equipment. It may also be used for local wiring in the switchboard. A good grade of switchboard cable is made up of tinned copper conductors coated with black enamel and silk and cotton insulation. The pairs are bound together with a spiral layer of paper, a layer of lead tape, another layer of paper, and a fireproof impregnated braid.

152

LOCATION OF SPLICES

Power cable is generally spliced in manholes or vaults, since it is used almost exclusively for underground distribution of power. Telephone cable is much less frequently spliced undei^ound, since it is most frequently used for aerial installations. Occasionally, advanced base construction might call for installation of underground telephone cable so as not to interfere with air operations and to improve the appearance of the base. However, most of the telephone cable you will splice will be aerial cable.

CABLE PREPARATION

After the cable is pulled into the rings or lashed to the supporting strand, the overlaps at the splice should be made secure from any longitudinal movement. A short length of bonding wire or ribbon soldered to the two pieces of cable will prevent the cable from creeping due to expansion and contraction caused by changes in temperature. This snubbing of the cable is especially important when there is a wide range of temperature or if an extended period of time will elapse between the ringing and the spUcing operations. If cable is not properly snubbed, cable bowing may result.

Open cable ends must be protected from moisture. If a splice is to be made within 24 hours the end may be protected with several layers of friction tape. If more than 24 hours will elapse before you make the splice, you must seal the ends by soldering.

To seal the end of a cable, bend it to an upright position so that you can work on it more efficiently. Then use a drift pin to drive the wires back into the sheath about)i to % inch. Remove the paper wrapping from inside the sheath; use a shave hook to clean the inside and the end of the sheath. Bevel the end and apply stearine to the cleaned surfaces. Use the cable dresser to dress in the edges of the sheath to form a roimded cup. Solder the end closed, using either a soldering iron and 50-50 solder or an acetylene or blowtorch and a stearine core solder. Float the solder until it is flush with the top, and then cool by applying stearine.

326605'—65 11 ' 53

A slack puller (fig. 6-1) can be used to obtain working space between the cable and the supporting strand. When the slack puller is removed, there will be no slack in the cable adjacent to the splice.

Fisure 6-1.—Slack puller installed between cable and supporting strand.

Recommended openings and sleeve dimensions useful in planning a splice are listed in appendix II. These are recommended dimensions and should be modified to meet prevaihng conditions.

REMOVING SHEATH

The first step in cable splicing is to remove the sheath. Measure and mark the length of sheath to be removed. The cable is ringed at the point marked. There are several tools available for ringing cable. Two such tools are shown in figure 6-2. Care must be used when working with the pUer-type ringing tool not to cut all the way through the sheath into the wires. The tool at the right in figure 6-2 may be adjusted to take a cut of varying depth. It should be set so that it will not cut all the way through the sheath. For cable larger than 1\'7d^ inches in diameter, use a chipping knife to ring and sht the sheath. After the cable is ringed, use the chipping knife to split the sheath longitudinally. Be careful not to cut into the insulation. After you have slit the cable, bend it back and forth at the ringed spot till it separates from the sheath.

PREPARING TO SPLICE

After the ringed and sUt portions of the sheath are removed from the cable, bind the cable to the sheath with two or three turns of 1-inch tape or muslin. Secure it firmly with a sHp tie. Tuck part of the binding under the sheath with a blunt tool. Now remove the paper tape and separate the wires according to color groups and tie them together with splicing-out wire or muslin. The standard cable color code is listed in appendix III. You will find it helpful to memorize this code if you do much telephone cable splicing.

Before you b^in to splice, boil out the insulation with hot parafi&n. This is done not only to remove moisture but also to make it easier to remove the paper insulation when splicing. The paraffin should be heated to a temperature of 375° F. Check the temperature with a special thermometer made for this purpose since paraffin applied to paper insulation at a temperature higher than 390° F. will seriously damage it. If such a thermometer is not available, you can judge the temperature by observing a dry ring that starts to creep up the outside of the pot at about 360° F. At the

proper working temperature this dry ring is about 2-inche8 wide at its widest point.

To boil out the cable, pour paraffin over the sheath, starting about 6 inches back from the exposed wires. Gradually work the stream of hot paraffin onto the paper insiiIatioQ and out to the end of the wires. Hot paraffin may be poured from a container with a spout and caught in a large shallow pan or it may be repeatedly ladled over the insulation from a deep container. When splicing lai^e cables, boil out the cable after splicing each 200 pairs; this removes moisture that the insulation may have absorbed from your hands or the atmosphere.

Instead of boiling out the splice with hot paraffin it may be enclosed in a muslin envelope and filled with a desiccant. A desiccant is a substance that absorbs moisture. If desiccant is used as a drying agent instead of hot paraffin, it should be applied to the splice from time to time to keep the insulation dried out. After the splice is completed, additional desiccant should be sprinkled into the completed splice.

Before the actual splicing begins, prepare a lead sleeve by cleaning it with the shave hook for a distance of 3 inches on each end. Coat the cleaned surfaces with stearine and slip the sleeve up over one of the cables to be spUced tiU it is out of the way. Wrap the cleaned portion of the sleeve with muslin if it will be some time before the spUce is completed.

After the sleeve has been slipped out of the way, fold the pairs back in both directions as shown in figure 6-3.

ELECTRICIAN'S SCISSORS

Now you are ready to begin the actual splicing of the pairs. Before going into this let's take a look at one of the most important tools used by a splicer—the electrician's scissors.

The electrician's or cable splicer's scissors, often referred to as shears or snips, are a short heavy bladed tool used to cut soft copper wire not larger than No. 18 AWG (American Wire Gage). The backs of the scissors are corrugated to aid

Figure 6-3.—Pairs laid back in color groups.

in removing the paper insulation of cable conductors.

The scissors are held in the hand with one finger through one of the loops in the handles as shown in figure 6-4. The

Figure 6-4.—Using electrician's scissors.

scissors are kept in the hand during the entire operation of joining the conductoi-s since too much time would be lost if they were laid down between individual steps in the operation. The way you hold the scissore is a matter of individual preference. Some pereons find it most convenient to place the fourth or little finger through the loop of the handle and operate-the scissors near the ends of the blades by manipulation of the thumb and first or forefinger; others prefer to use

the third or ring finger in the loop of the scissors and manipulate them with the thumb near the pivot screw. This method of holding the scissors and cutting is shown in figure 6-4. Hold the scissors however it is most convenient for you, but never attempt to work with your thumb through a loop in the handle.

The telephone cable splicer and his helper usually carry their scissors with them whether they are actually splicing cable or not. Their scissors are the badge of their craft, and rightly so, since they use them more than any other tool in their kit. Since scissors are always at hand the inexperienced splicer has a tendency to use them for all types of jobs for which they are not intended. It is a good rule to use the scissors only for cutting small-gage wires not larger than No. 18 AWG or for skinning wrapping paper and pulp insulation. Scissors should never be used as a tucking tool when butting under the cable sheath with cotton tape. Use a small blunt screwdriver.

Fisure 6-5.—Electrician's tciston pouch.

158

Splicing scissors do not operate properly if the pivot screw is allowed to become loose. The screw should be tightened whenever necessary with the proper size screwdriver, not by battering with a hammer or squeezing in a vise.

Don't carry splicing scissors loose in your pockets, as they can inflict serious injuries. Carry them in a scissors pouch as illustrated in figure 6-5. If no pouch is available through supply channels, make one out of leather or canvas.

CLASSIFICATION OF SPLICING

In general there are three types of telephone cable splicing as illustrated in figure 6-6.

C

Figure 6-6,—A, straight splicing; B, bridge splicing; C, butt splicing.

Straight splicing is the process of joining two wires which are at opposite ends of a splice. Bridge splicing is the process of joining three or more wires together to form a bridge. Butt splicing is the process of joining two wires which enter the splice from the same end. A butt splice m&y be made with a butt joint, or the wires may be bent over and made into a straight joint. Butt splices are made for flexible arrangement of facilities; they may later be changed to straight or bridge splices.

JOINING THE PAIRS

If you watch an experienced cable splicer you will soon learn that he moves quickly with a minimum of waste motion. His speed is the result of practice and following an established procedure. The procedure for joining pairs that follows may not be the exact one you will use when you are an expert cable splicer, but it is a good place to begin.

Let's follow step by step the procedure for splicing a single pair of conductors. Figures 6-7 through figure 6-21 illustrate each operation. Refer to these figures as you read the description.

1. Grasp a pair of wires on the right-hand side of the splice with the left hand. At the same time cross the right hand over the left hand and grasp a pair of wires on the left-hand side of the splice. Grasp the pairs between thumb and forefinger.

Fisure 6-7.—Operation 1: Selecting pairs. 160

Fisure 6-8.—Operation 2: Bringins wires together.

2. Bring the wires together, leaving slack on the far side of the splice. Gradually diminish the slack as each pair is spliced until the center or core of the cable is reached to ensure a sjmmetrical splice.

Figure 6-9.—Operation 3: Twisting pairs together.

3. Hold the wires together and give them a sharp half turn.

4. Take both pairs of wires in your left hand, and slide your hand along the wires to hold them while they are being cut. Cut the wires approximately 4% inches from the twist.

5. Grasp the wires just below the twist and pull the insulation off the ends of the wires. Pull awarrfrom the twist and slightl\'7d'^ toward the side on which the cotton sleeves are to be slipped. This prevents the ends of these wires from curling.

Rgur* 6-10.—Operation 4: Cuttins wires to length.

6. "While the left hand is stripping the insulation, take a pair of cotton sleeves in the right hand and hold them parallel to one another, but with their ends staggered approximately % inch.

7. Hold the pair of wires on the left side of the splice in the left hand and slip the sleeves over them with the right hand.

8. Slide the sleeves along the wires with a rolling motion until they have reached a position where they do not interfere with twisting the joints.

9. When the sleeves are in position and do not interfere with twisting the joint, bend one of the wires back so it is out of the way while the other wires are being spliced.

162

Fijurc 6-12.—Operation 6: Prcparins to ploct slc«v«t.

Figure 6-14.—Operation 8: Sliding sleeves info place.

10. Draw the wires together so that the twist can be made close to the ends of the insulation.

11. Grasp the wire between the thumb and forefinger of the left hand, roll the wire on the left-hand side over the wire on the right-hand side. Take the wires between the thumb and the forefinger of the right hand and bend them at right angles. Hold the wires tightly in the left hand and twist them by crank-handling with a wrist movement. Using the wrist is much less tiresome than twisting with the entire arm. The crank should be about J^-inch long.

Fisurc 6-17.—Operation 11: B«3innin3 of the crank handle twist.

12. Continue crank-handling until the twist is forced beyond the thumb and forefinger of the left hand as shown in figure 6-18. After you acquire some skill you can make the twist by giving the crank three turns with the wires held tightly in the left hand followed by three turns with the wires held loosely.

Fisure 6-18.—Operation 12: Completion of twitting of wires tosethtr.

13. After crank-handling, the joint should contain the number of tAvists indicated in figure 6-19.

14. Slide the left hand along the pigtail, away from the neck of the twist to serve as a

guide for the shears; then grasp the sleeve with the left hand. Bend the pigtail with the shears at an angle of 45° to the wire and cut the excess twisted wire off squarely, at the same time the sleeve is slipped over the neck of the twist.

Figure 6-19.—Operation 13: Inspection for proper number of turns in splice.

Figure 6-20.—Operation 14: Trimmins the joint.

15. While you are slipping the sleeve over the twist, reach for the second wire with the right hand. When the sleeve is in place, grasp the remaining wire with the left hand. Make the twist as for the first wire. Continue in the same manner to join all the pairs making up the splice.

Fisurc 6-21.—Operation 15: Slippinj the sleeve over Joint.

This procedure may be varied by placing the sleeves on a number of pairs before beginning the splice. The sleeves are slipped on the wires and the pair is given two or tliree crank-handle turns to hold the sleeves in place till the splice is made.

BRIDGE SPLICING

Basic splicing techniques have been illustrated in describing the procedures used to make a straight splice. As mentioned previously, bridge splicing is the process of joining three or more wires together to form a bridge.

When a splice exists at the point where the branch cable is to be bridged, twist the joint by slipping the sleeves back one at a time to uncover the joint. Skin the insulation off of the branch pair to expose 6 inches of bare wire. Now wrap the wire of the branch cable tightly around the pigtail of the existing joint. Give all three wires of the pigtail a full turn and cut off % inch of the pigtail. Slip the sleeve back over the joint to complete the splice.

When there is no splice in the main cable where the branch is to be bridged, twist the joints using this procedure: First cut the pairs. Then skin all three pairs of wires to allow 3 inches of bare wire. Now take a piece of insulated wire 12-inches long and of the same gage as the main cable and join it to one of the wires of the main cable with a

twisted joint. Trim the joint, bend the pigtail down, and slip a sleeve over it. Now slip a sleeve on the other wire of the main cable pair. Twist the two ends of the main cable wires and the bridge wire into a straight twisted joint. Bend the pigtail down and slip the sleeve over. Follow the same procedure for the other wires of the pair and the other pairs in the splice.

DRYING THE SPLICE

After the wires are joined, the splice must be dried. This can be accomplislied hy boiling out the insulation with hot melted paraffin or by the use of a desiccant.

Although the use of hot melted paraffin is effective and finds application in many splices, it has a number of disadvantages. Melting and heating paraffin create a fire hazard arid present a burn hazard for men working on the cable. Paraffin heated inside a building gives off objectionable fumes and may damage walls and floors if splashed on them.

The use of a desiccant overcomes these disadvantages. The action of a desiccant in a cable joint may be compared to that of a sponge sucking up water. Some splices treated with desiccant quickly attain high insulation resistance between conductors; others require days or even weeks to attain high insulation resistance. The length of time for a splice to dry depends on the amount of moisture present, size of splice, tightness of binding, and type of insulation. The size of the splice affects drying time; small splices usually dry more rapidly than large ones. Tight binding of splices tends to slow up the drying rate. Paraffined sleeves and insulation give up their moisture more slowly than unwaxed ones.

Anhydrous calcium sulphate and silica gel are the principal desiccants used in drying splices. Anhydrous calcium sulphate consists of irregular white granules approximately)/i6-inch across. Silica gel is a silica compound issued in the form of tiny particles similar to white sand. Anhydrous calcium sulphate wiU absorb a little over 6 percent of its

weight of water and still have zero vapor pressure. If a pound of this desiccant were sealed in a splice containing an ounce of water, all of the water would be absorbed.

To apply desiccant, envelop the splice with a single piece of muslin 2 inches longer than the splice opening and overlapping the splice at least 1 inch at the top. Open the overlap at the top of the splice and spread the conductors apart with your fingers. Now pour the desiccant granules into the spaces between the conductors. Be sure there is ample distribution in the ends of the muslin cradle near the sheath butts. Close the muslin cradle and wrap the entire splice in two layers of unboiled muslin bandage.

After a can of desiccant has been opened, any material left in the can at the end of the day should be thrown away to make sure that no moisture-saturated desiccant is used in a splice.

The amount of desiccant required varies with the type and size of the cable. Loose-core, heavy^age, double-wrapped cables require approximately one gram per pair. Small-gage, single-wrapped cables such as 22- and 24-gage exchange cables take approximately % gram per pair. For example, a 200-pair, 16-gage, double-wrapped cable would require approximately 200 grams, A 22-gage, 100-pair, single-wrapped cable would require about 20 grams. (One ounce is equal to 28 grams.)

WIPING THE SLEEVE

After the splice is completed and treated with desiccant or boiled out with paraffin, you are ready to begin the sleeve joint. The sleeve, which you have previously slipped aside, is now ready to be centered over the joint.

Use a standard cable dresser to beat in the ends of the sleeve before wiping; this makes a tight joint. The beat-in portion of the lead sleeve on straight joints should be rounded and full. Revolve the sleeve with one hand and hold the cable dresser in the other. Using quick, sharp strokes, tap the end of the sleeve to center it around the cable, free from dents and furrows. Furrowing or petticoating is prevented

326605»—5«5 12 1*9

by distributing the beat-in j>ortion evenly over the surface of the sleeve that is bent in to meet the cable. While you are beating in the sleeve be very careful not to strike the cable sheath. The final tightening of the sleeve can be done with a riveting hammer. Don't strike the cable with the hanmier.

After the sleeve is beat in, place the pasters (gunmied paper) as shown in figure 6-22.

PASTERS

Rjurc 6-22.—Fasten in plac*.

Now the joint is ready to be wiped. Good wiped joints are difficult to make and it requires a great deal of practice and experience to turn them out consistently.

To begin with you must have wiping metal of the right composition. It should be approximately 40 percent tin and 60 percent lead. Some splicers prefer slightly more than 60 percent lead so they add a small amount of pure lead to the 60 percent, 40 percent mixture furnished by the manufacturer. A wiped joint that tends to run at the bottom when it is hot indicates a high tin content. The addition of a small amount of lead will usually remedy this situation. If, on the other hand, the percentage of lead is too high, the joint may wipe easily but it will present a chalky appearance and may be porous. This condition can be corrected by adding 50-50 solder to the wiping metal.

The wiping metal should be used at a temperature of 670° F, Check the temperature with a thermometer. If no thermometer is available, you can check the temperature by folding a piece of dry paper and plunging the end into the hot

metal and removing it immediately. If the paper is scorched to a point where it is about ready to ignite, then the metal is ready to use.

Before starting to wipe one end of the sleeve, tack the opposite end by pouring a few drops of solder on it.

In wiping the joint there are six distinct steps: tinning, heating (sometimes referred to as building up the heat), rough forming, wiping (or finishing), cooling, and inspection and removal of pasters.

When first pouring the solder, splash it onto the pasters; at the same time move the catch cloth back and forth under the joint. At the start of the operation, do not pour solder directly on the cable sheath or sleeve as it might burn through and damage the conductors. Continue making a circular motion with the ladle, splashing solder on the pasters and joint untU the entire surface to be wiped is weU tinned and no dark spots appear on the sheath or sleeve. During the tinning operation, do not attempt to bring the solder in the catch cloth up on the sides or top of the wipe. To do so would probably result in a torn cloth, since sharp points form on the chilled metal. The purpose of the catch cloth during tinning is to ensure that the bottom of the joint is tinned and that- the solder does not form a hard lump on the bottom of the wipe. This explains why the cloth is moved back and forth. Many men, before they gain experience, tend to hold the catch cloth too close to the cable and sleeve. It should be held at a distance of at least % inch to prevent solder from freezing on the bottom.

After the entire area is tinned, the next step is to heat the joint. If the sheath and sleeve are not heated to the proper temperature (approximately that of the wiping solder) it will be impossible to turn out well-finished wipes. To heat the joint, continue to pour metal onto the joint until it runs freely from the joint. During the pour, bring the metal in the catch cloth up to the top of the splice from both sides and then work it into the ladle. If the metal is chilled, it should be returned to the solder pot and a fresh ladleful taken. If the metal is still reasonably hot, some of it can be restored

to the pot while the part remaining in the ladle is supplemented by fresh solder from the pot. The addition of stearine to a joint dm*ing the heating process helps to make the metal handle better. A poorly heated joint will cause small mounds of chilled metal to appear at the paster edges. Continue heating until these mounds are no longer present. Protect the wipe with the catch cloth while the ladle is being refilled so that the heat will be maintained.

When it is apparent that the joint is properly heated, it is time to begin the rough forming. You will learn from practice and experience when the mass of solder has reached the right stage for rough forming. At this stage, the mass is pasty but is still quite fluid and the tin content of the solder tends to run toward the bottom of the joint. During rough forming place the ladle in the solder pot or drip pan. Manipulate the mass and bring the excess metal to the top of the joint. The process of forming consists of a rapid packing motion by means of the catch pad and a finishing cloth. At the same time, the metal aroimd the joint is shaped or built up by the action of the cloths. This packing should not be continued for too long or the chilling of the metal will make it impossible to finish or wipe the joint successfully.

The joint should be finished with as few motions as possible. Three motions are considered the ideal number. The closer you approach this ideal minimum the better your wipes will be. Repeated wipes in finishing a joint result in a job that is rough and chalky in appearance. Such a joint is commonly referred to as a cold joint, because it looks as though it had been wiped with cold metal. When preparing to finish the joint, hold the finishing cloth at the top edge between the thumb and forefinger of each hand, with the middle fingers alongside the edges to exert pressure. Bring the first wipe from the bottom to the top of the joint on the side away from you. Perform the second wipe in the same manner without altering the position of your hands. This enables you to see the condition of the bottom of the joint during the final wipe and allows you to maintain equal pressure on both sides of the joint while completing the wipe. It is not necessary to

move your feet at all in finishing a wipe. This is especially important when wiping aerial cable, since the slightest movement on the aerial splicing platform may result in a cracked splice. The third wipe should be a light longitudinal stroke parallel with the sleeve and cable. Its purpose is to remove the ridge on top caused by the first and second wipes.

Allow all fresh wipes to cool enough so that slight movement of the joint will not cause cracks.

After the joint has cooled, inspect it for defects. If any defects are found, rewipe the splice. No attempt should be made to cover up a defective splice by trinmiing with the splicer shears or removing roughness with a knife, rasp, or hot ladle edge. Pinholes should never be treated with a soldering iron or hot screwdriver. To do so would only defeat the purpose of the wipe and invite future breakdowns and other trouble. Don't touch up a wiped joint with a rasp file or emery cloth to improve its appearance. The finished wipe should have a good appearance because it is properly made, not because some questionable corrective measure has been applied.

SPLICING SILK- AND COHON-INSULATED CABLE

Paper-insulated cables are not used for direct termination to the springs or punchings on central office distributing frames or terminal strips. Paper insulation is not used, because paper-wrap or paper-pulp insulation is not strong enough to stand the handling necessary to make the core into a form which is fanned out and laced. In addition, paper insulation deteriorates rapidly when exposed to the air. Hun\idity soon lowers the resistance of the paper insulation and trouble is encountered.

To overcome these disadvantages of paper insulation, short lengths of lead-covered, textile-insulated cables are made up into forms, called cable heads, to be connected to the distributing frames, terminal strips, and other interior terminal points. Cable heads are long enough to permit splicing the unformed end to the incoming paper-insulated cable.

Silk- and cotton-insulated cables are prepared for splicing by first being boiled out in beeswax or a special petroleum wax. Paraffin is not used for boiling out silk- and cottOD-insulation because it will discolor the textile insulation and make it difficult to separate the conductors for fanning, forming, and sphcing.

Plastic-insulated cable requires no boiling out.

Do not use beeswax for boiling out paper-wrapped or paper-pulp insulation; it makes the paper brittle and may cause breakage.

The same standard spUce openings, number of cotton sleeve banks, and splicing procedures apply to textile-insulated exchange cables as apply to paper-insulated exchange cables.

To mark the point for removal of textile insulation, use the same procedure as used for paper-insulated cables. Cut off the conductors 4% inches from the end and use long-nosed pliers to crush the silk- and cotton-insulation at the marked point. Don't apply too much pressiu'e as this will flatten the conductor and may cause it to break after it is spliced. After the insulation has been crushed, it can be removed by decreasing the pressure on the pUers and pulling toward the end in a straight line. The insulation can also be pulled from the conductors with the fingers. Never use the back of the splicing scissors to remove the insulation as you do when dry splicing paper-insulated cables. Such a method might break the relatively brittle tinned copper conductors in textile-insulated cables.

Some textile-insulated cables have enameled conductors. This enamel must be thoroughly removed before splicing the wires. One way to remove it is by lightly scraping the conductor between the jaws of long-nosed pliers. You must be extremely careful not to nick or flatten the wire. A good tool for scraping enamel is the wire scraper, which consists of a steel spring ^-inch wide, bent in the form of a U with the ends turned and sharpened. As you gain experience with this tool, you will find it effective for removing silk, cotton, or plastic insulation.

To twist the conductors, join them and give the pigtail a 3- to 5-tum twist as you would in splicing paper-insulated cables. Be sure to include about K inch of insulation in the neck of the twist to prevent snapping the conductors at the point where the insulation was removed and to hold the cotton sleeve in place more firmly.

After the wire work on a silk- and-cotton to paper splice is completed, boil 'out the splice in paraffin and wrap as for a paper-insulated splice. A desiccant may be used instead of the boiling out process.

CABLE REPAIR

Making minor repairs because of ring cuts, sheath breaks, lightning bums, bullet holes, and other damage to small aerial cables may be part of your job. Such damage can be repaired by the carbon electrode welding method or with an acetylene torch. The carbon electrode method is easier to use if damage extends through the sheath and moisture has reached the conductor insulation. The acetylene torch method is preferable if the damage does not extend through the sheath.

Water-soaked paper insulation usually causes the pairs to become shorted. Shorted pairs are frequently the first indication that a sheath has been damaged so much that moisture can enter.

Use the following steps for repairs of this nature:
1. Open and spread the sheath.
2. Dry out the insulation with desiccant.
3. Repair charred or corroded insulation.
4. Test the cable pairs before preparing to close the sheath.
5. Close the sheath.
6. Solder the seam and restore cable hangers to appropriate spacings.

OPENING THE SHEATH

After the fault has been located and the ladder or platform has been placed, proceed as follows to open the sheath:

I I

I

1. Straighten the cable if necessary. If the cable is under tension, lessen it by placing grade clamps to hold the cable ! as desired until the repair operation has been completed.

If the cable is not extremely taut, this wiU not be necessary.

2. With a wire brush brighten the lead sheath in the section to be split.

3. Rub stearine along the cable where the longitudinal cut is to be made. As the cut is made continue to lubricate with stearine.

4. Adjust the small blade of the cable stripper so that it will penetiate jit to K« inch. Make the starting cuts with the small end of the blade. Rout out the small shavings of lead. Each cut will make the succeeding cut easier. After a definite line of cut has been established, turn the tool over and use the large end of the blade to bevel the cut. Don't cut all the way through the sheath with the lai^e end. Just before the blade is about to break through the sheath, reverse the tool and complete the cut with the small end.

5. After the sheath is cut through, open it with cable pliers. Grasp the handles of the cable pliers firmly and open the incision on one side. Place the sharp jaw in the cut and work the tongs sideways, exerting a small downward pressure at the same time. Bend the sheath cautiously so as not to put too much stress on it and to avoid deforming it. When one side of the cable sheath is bent back, repeat the process on the other side. After the cable sheath is open, push out the paper-insulated cable by inserting a wooden wedge between the sheath and the wires in the manner shown in figure 6-23. Insert enough wedges between the core and the sheath so that you have room to work. Remove the paper wrapping to expose the wires so that they can be dried out by using a desiccant or by boiling out with paraffin.

REPAIRING INSULATION

If it is necessary to splice out the wires, stagger the joints and scatter them along the length of the opened cable. If necessary, lengthen the cut to permit this. Use paper tape to avoid piecing out wires on which the insulation has

WOOD WEDGES

3

Figure 6-83.—Wcdsct insctltd in cable to permit dryinj out.

become blackened because of electrolytic action. To apply the tape pull the pair out and cut a piece of tape long enough to extend at least K inch beyond the corroded portion. Press it down with your fingers. Trim off the excess tape with your scissors. In hot weather take care that perspiration from your hands does not dampen the insulation.

After all damaged wires are repaired, push cotton tape under the edges of the sheath with a butting tool. Wrap the core with 1-inch tape which has previously been boiled out in paraffin and push the core back into the sheath. Test the cable pairs to be sure that all trouble is cleared. Before closing the cut in the sheath, cut a piece of press board about Ke-inch wide and equal to the length of the opening. Lubricate the top of the press board generously with stearine so that the edges of the cut will slide back into their original position more readily. Use tongs to press the lead sheath back into its original position. It is not necessary to close the cut entirely, but if the cut is left partly open, the press board should cover the core so that it will not be damaged if the electric welding process is used in closing the cable.

In closing the sheath, first squeeze it together as much as possible with your hands. Shape the cable into its original circular shape by applying pressure around it with the tongs. Do not apply pressure directly over the cut until the cable is well shaped. To do so may flatten the cable. The inside of the tongs should be lubricated frequently with stearine. Do not roll the tongs over the sheath, as this will produce depres-sions around the circumference. Work the cable into shape gradually.

For welding the seam, you will use the carbon electrode lead burning outfit described in connection with battery repair in Construction Electrician*8 Mate S (& NavPere 10636-A.

After the sheath has been closed over the core so that the edges of the cut are from Ys to % inch apart, use the scoring tool and a wire brush to clean the sheath. Apply pasters along the edges and ends of the cut so that the width and length of the closing joint will be fixed. Place the pasters about Hs to \'7di inch from the edge of the cut. Use a half-round bastard file at each end of the cut to form a depression in which the solder can be flowed to seal any fine cracks which may have been caused by bending the sheath back. After the pasters are in place, apply stearine to the portion to be soldered.

Stearine core solder should be fused to the sheath by holding the end of the solder in contact with the sheath and touching the solder with the edge of the beveled end of the electrode. In this manner coat the entire area to be soldered. Use only the minimum amount of solder. Heat and manipulate the solder deposited on the sheath with the beveled surface of the electrode to tin the entire area of the patch. To ensure the tinning of the entire surface of the deep narrow cut, turn the electrode so the edge of the beveled surface can be moved along inside the cut. Light contact and a circular motion of the electrode on the solder will aid in raising the , temperature of the solder so that the tinning can be accom- , plished to best advantage. Additional solder should be flowed into the cut to fill it to the top. Move the electrode \ over the patch to remove air bubbles and flux and to give the repair a reasonably smooth appearance. The completed | patch should project very little above the surface of the sheath. After completing the welding, remove the pasters | and use a coarse file to smooth off any rough spots in the ^ seam. Smooth off the sheath with a ^^'i^e brush. Leave the cable as sound and as free from marks or creases as possible.

SHEATH REPAIR WITH TORCH

As previously mentioned, ring cuts and small defects that do not extend through the sheath may be repaired by using an acetylene torch. This method is particularly effective for cracks that occur while the cable is being worked. The acetylene torch can also be used to repair cracks extending through the sheath, but care must be taken not to damage the insulation.

Acetylene is a highly flammable gas and is explosive in an enclosed space. Never use an acetylene torch in a tent, manhole, underground, or similar confined spaces. Also, be sure that you fully understand the operation of an acetylene torch before attempting to use one. The Navy Training Courses Steelworker S <& 2, Pipefitter S <Ss 2, and Metalswith 3 i& 2, &s well as various commercial publications, explain the principles of operation, including safety precautions.

Stearine core solder should be used to make sheath repairs with the acetylene torch. Thoroughly clean the area around the defect with a brushing motion. Apply the flame to the cleaned area, being careful not to concentrate the flame in one spot. Don't allow the blue cone in the flame to come into contact with the sheath. Apply solder over the area of the defect, using sufficient heat to tin the area thoroughly. Build up enough solder to fill the cut or depression. Use a small finishing cloth to pack the solder into the defect and smooth off the patch so that it is only slightly above the level of the adjacent sheath. The repairs should extend from % to % inch beyond the defect.

You may make adjacent repairs without tiuiiing off the gas in the cylinder. However, if you change position or pass a pole, extinguish the torch by turning the needle valve. Never keep the torch lighted except when making repairs.

TRICKS OF THE TRADE

Telephone cable splicing is a complex operation requiring a great deal of skill and experience. The discussion in this chapter has necessarily been limited to a few of the fundamental principles. You will learn "tricks of the trade" and shortcuts only by experience. However, you can pick up many pointers from the Army technical manual Lead-Sheath Telephone Cable Splicing (TM 11-372), from trade publications of the American Telephone and Telegraph Co., and perhaps from other sources. Even after you have gained considerable experience, you will find many helpful hints on splicing in technical publications.

QUIZ

1. With what material is telephone cable wire usually insulate<17

2. What is high dielectric insulation?

3. In preparing cable for splicing, what method is used to prevent cable creeping due to changes in temperature?

4. What is the maximum time cable ends may be left open without being soldered?

5. To solder the end of a cable closed, what equipment is used?

6. In splicing large cable, how frequently should the cable be boUed out?

7. Name the only purposes for which electrician's scissors should be used.

8. What is bridge splicing?

9. What methods are used to dry a splice after the wires are joined?

10. When a desiccant is used, what factors afTect the length of time required for a splice to dry?

11. Name the principal desiccants used in drying splices.

12. What should be the composition of wiping metal?

13. Name the six steps in wiping a joint.

14. In wiping a joint, what is the eflfect when the joint is poorly heated? 16. What is a cold joint?

16. How can the cooling process of a wi;)ed joint be hastened?

17. What material is used to boil out silk- and cotton-insulated cables being prepared for splicing?

18. In splitting cable to make minor repairs, to what depth should the small blade of the cable stripper penetrate?

19. How far should repairs made with an acetylene torch and stearine core solder extend beyond the edges of the defect?

20. What Army technical manual contains information on telephone cable splicing?

CHAPTER 7

TRANSFORMER INSTALLATION AND REPAIR

YOUR JOB

If a blueprint of a particular transformer installation is available to you, then your job will be comparatively easy. All the construction and electrical specifications will be worked out for you beforehand and all you have to do is convert this information into the finished job. However, in some instances, a blueprint will not be available. Then it will be up to you to:

1. Determine the location and size of the transformer.

2. Install the transformer according to the fundamental standards of the National Electrical Code.

DETERMINING TRANSFORMER SIZE

You've been given the job of installing a single-phase transformer in a certain area of the advanced base. This area contains 10 buildings that receive power from a 2,400-volt overhead primary main. The electrical equipment in the buildings consists of single-phase lights or motors operating at either 110 or 220 volts. A three-wire overhead secondary main distributes the secondary voltage alongside the buildings. Service leads complete the connection between the secondary main and each building.

The first thing you should do is make a rough drawing of the area. When you get done it should look like the sketch in figure 7-1. The location of each pole as well as the buildings is noted. Lines, representing the service leads, are drawn between the poles and the buildings.

Your next step is to determine the maximum power demand of each service. It sounds complicated, but what it actually amounts to is summing up the power required by the lights and motors in each building. This jx)wer demand is noted in each square representing a building (fig. 7-1).

Next, figure out the kv.-a. load per pole. In this particular example, each pole serves two buildings. Therefore, the kv.-a. load of a pole will be the sum of the maximum power demands of the two buildings served by that pole.

Now, calculate the total maximum connedted load on the transformer. As you can see from figure 7-1, the total connected load is the sum of the kv.-a. loads per pole. It amounts to 25.85 kv.-a. But don't jump to conclusions. You won't need a 37.5 kv.-a. transformer to take care of the total load. 25.85 kv.-a. represents the amount of power that the transformer would have to supply if all the lights and motors were turned on at the same time. Although that possibility exists, the time interval would be very small compared to the length of time that only a portion of the total load would be on. Therefore it is only necessary to calculate an average demand and then use this figure as a basis for determining transformer size.

POLES
KVA LOAD PER POLE
BLD6. A .5 KVA
BLDG. C 2.7 KVA
BLDG. E .15 KVA

<

o q:
BLDG. D
1.3 KVA I ' I 4.0 KVA
BLDG. F _ , , 2.0 KVA m t 2.16 KVA
BLDG. G 5 KVA

BLDG. H 1.9 KVA 1§ ^ 6.90 KVA
BLDG I 4.2 KVA

i
BLDG. J _. 7.8 KVA I; . W 12.0 KVA
STEP I TOTAL CONNECTED LOAD = .8 + 4.0+2.15+6.9+I2.0» 25.85 KVA.
STEP 2 AVERAGE DEMANDS 25.85 X .60 « 15.51 KVA. STEP 3 TRANSFORMER SIZE REQUIRED-25 KVA.
Figure 7-1.—Transformer calculations. -

Average demand can be approximately computed by multiplying the total maximum connected load by 60 percent. In this example, the average demand is 15.51 kv.-a. (25.85X0.60). The transformer capacity required to meet this demand will be 25 kv-.a., since a 25 kv.-a.

transformer is tlie next largest standard size.

DETERMINING TRANSFORMER LOCATION

Your next problem is to find the most suitable location for the transformer. That doesn't mean finding the strongest pole, but the one that is nearest to the electrical center of the area.

The electrical center is the point where a balance is obtained between the total kv.-a. spans to the left and right of the transformer's location. The kv.-a. span is the product of the number of spans times the kv.-a. load of the pole.

To begin with, assume that you are going to place the transformer on pole K (fig. 7-1). Then, figure the total kv.-a. spans to the left and right of this location. A chart will simpHfy your calculations—

Kv.-a. spans to left of imie K Kv.-a. spans to rieht of oole K

1X4. 0=4. 0 IX 6. 9= 6. 9

2X . 8=1. 6 2X12. 0=24. 0

5. 6 30. 9

Total kv.-a. spans Total kv.-a. spans

You can see that if you placed the transformer on pole K, it would be at an unbalanced electrical center. That is, it would be too far away from the heaviest loads. So pick another pole. This time choose pole L and make another chart—

Kv.-a. spans to left of pole L Kv.-n. spans to rij?ht of pole L

1X2. 15= 2. 15 1 X 12. 0= 12. 0

2X4. 0 = 8. 0

3X . 8 = 2. 4 12.0

Pole L is nearest to the electrical center of the area. That's the pole, then, on which you will mount the transformer.

A typical single-phase transformer installation is shown in figure 7-2. The primary voltage, carried on a three-wire

to left of pole K=5. 6

to right of pole K=30. 9

12. 55

Total kv.-a. spans

to left of pole L= 12. 55

Total kv.-a. spans

to right of pole L = 12.0

SINGLE-PHASE TRANSFORMER INSTALLATION

POSITION OF PHASE CONDUCTOR WHEN NO SfKIES STREET LIGHTING CIRCUIT IS TO BE INSTALLED, IF SERIES CIRUtT IS INSTAllED, IT MUST BE IN THIS POSITION AND PHASE CONDUCTOR SHIFTED TO POLE PIN POSITION

FUSED CUTOUT

NOTE:

PRIMARY CONNECTIONS BETWEEN UNE CONDUCTOR AND TRANSFORMER BUSHING MUST BE WP INSUIATEO NO 6 COPPER

- HOT-LINE CLAMP OR SOLOERLESS CONNECTOR

COMMUNICATION SYSTEM 20'-0" MIN TO GROUND

FRONT VIEW SIDE VIEW

Figure 7-2.—Single-phase transformer installation.

delta primary main, is stepped down throug:h the transformer to a three-wire secondary main. The actual installation of the transformer and its component parts is purely a mechanical process that you probably have done many times. However, as a review, check the following steps:

1. Bore two ^s-inch holes in a vertical line at the point where the transformer is to be mounted on the pole. The spacing of the holes will be determined by the spacing of the support lugs on the transformer tank.

2. Cut a gain on each end of the top hole and on the transformer side of the bottom hole.

3. Drive a %-mch machine bolt into each of the two holes. Make sure that you insert a flat washer against the gains on the transformer side of the pole.

4. Hoist the crossarm and guide it onto the end of the top machine bolt, tighten the nut, leaving enough space for the support lug of the transformer to slip onto the head-end of the machine bolt. Do the same with the lower machine bolt.

5. Aline the crossarm and fasten its braces to the pole by means of a lag screw. Make sure the braces are tightly secured to the crossarm.

6. Hoist the transformer and guide its support lugs onto the head-end of the machine bolts. Swing around to the other side of the pole, and draw up the nut on each machine bolt.

7. Mount the lightning arrester and fused cutout near each end of the crossarm. In figure 7-2, a combination arrester and cutout bracket is used that may be slipped onto and clamped securely to the crossarm.

8. Train a connecting wire from each primary bushing to the bottom of each fused cutout. Use the pin insulators as support for this wire.

9. Train a connecting wire from the top of each fused cutout to each main primary wire. Solderless connectors are used to join the connecting wire tiO the main wire.

10. Connect a wire between the top of each lightning arrester and the top of each fused cutout.

11. Connect a wire from each of the three secondary bushing terminals to their respective wire in the secondary main. Solderless connectors are again used.

12. Train a wire from the bottom of each lightning arrester, under the crossarm, and make a solid interconnection to the ground wire on the side of the pole.

13. Loop a wire from the neutral wire of the secondary main over to the ground wire and make a solid interconnection.

All this work will be done while the primary main is deener-gized.

SINGLE-PHASE TRANSFORMER INSTALLATION REQUIREMENTS

When a transformer is installed it will be up to you to see that the job is done right. That means you should be up to snuff on the rules and requirements of the electrical code. The code books that cover these rules and requirements are hsted in chapter 2. Due to lack of space, it would be impossible to reproduce all of the regulations found in the code books. However, a few of the more important transformer installation rules are listed here for your benefh:

1. Single-phase transformers larger than 100 kv.-a. must be installed on the ground or on a platform.

2. Single-phase transformers less than 100 kv.-a. may be mounted directly to the pole if equipped with support lugs or installed on crossarms with hangers.

3. Single-phase transformers of 50 kv.-a. or smaller are placed ABOVE the secondary mains if conditions permit. Those lai^er than 50 kv.-a. are placed below the secondary mains.

4. Lightning arresters and fused cutouts must be installed on the primary side of all distribution transformers except the self-protected type.

5. Ground wires must be covered with a wood moulding to a point 8 feet above the base of the polo.

The rating (size) of the fuse link used in the primary cutout is also important. If you choose a fuse size that is too small, needless outages will occur. The chart in figure 7-3 will help you pick the correct fuse size. As an example, suppose you have installed a single-phase 75 kv.-a. transformer that operates from a 2,400 volt primary. Using the chart, first find the transformer capacity (75) on the ordinate. Then proceed horizontally imtil you intersect the primary voltage hne (2,400 volts). The point of intersection lies in the 100-ampere fuse-link area. Therefore, for this particular installation you would use a 100-ampere fuse.

The size of the training wires used to make the connections between the transformer bushings and the primary and secondary mains is also an important consideration. In

fiSwt 7-3.—Fuse sizes for single-phase installations.

general, number 6 weatherproof insulated copper wire, or equivalent, will prove satisfactory as training wii-e for the primary connections of transformers, in sizes up to 200 kv.-a. The size of the secondary training wire, however, will vary with the size of the transformer as shown below in table 1. In all cases, the secondary training wii-e should be weatherproof.

Table 1.—Proper size secondary training wire for transformer

THREE-PHASE TRANSFORMER BANKS

In the majority of cases, you will employ a bank of interconnected single-phase , transformers to transform three-phase power from one voltage to another. The use of single-phase transformers provides a flexible arrangement that enables you to obtain either wye (star) or delta connections of the primary and secondary windings.

The advantages and disadvantages of the common types of three-phase transformer bank connections are fully discussed in chapter 6 of Construction Electricians Mate 3 and 2, NavPers 10636-A. Therefore, they will not be repeated here.

The total amount of power that can be supplied from three single-phase transformers connected in a three-phase bank is the sum of each transformer's kv.-a. capacity. Three 25-kv.-a. transformers, for example, can be expected to supply 75 kv.-a. of power under stated cooling conditions, provided that the transformers are approximately equally loaded.

The protective requirements of a three-phase transformer installation is the same as that of a single-phase service. That is, a lightning arrester and fused-cutout is required in each phase wire of the primary main feeding the transformer. The size of the fuse, again, is determined by the total capacity of the transformer bank and the value of the primary voltage. The chart in figure 7-4 will help you select the proper size of fuse. As an example, suppose that the transformer bank capacity is 75 kv.-a. and operates from a 2,400 volt primary main. You start at the 75 mark on the ordinate of the chart and proceed horizontally until you intersect the 2,400-volt line. This intersection lies in the 60-ampere fuse area; therefore you would use a 60-ampere size fuse.

THREE-PHASE TRANSFORMER BANK INSTALLATION
(POLE)

Three single-phase transformers, each rated less than 15 kv.-a., may be mounted on a single pole for a three-phase

Figure 7-4.—Fuze sizes for three-phase installations.

installation. The transformers can be secured directly to the pole if equipped with mounting; lugs or hung on a double crossarm if equipped with hangers.

Figure 7-5 gives an over-all view of a typical three-phase installation using tliree single-phase transformers. Each transformer has a capacity of 3 kv.-a. and as a bank, are rated at 2,400/4,160 volts on the high-voltage winding and 120/240 volts on the low-voltage winding. The3^ conform to NEMA standards and therefore have their winding leads arranged for additive polarity. The high-voltage windings are wye-connected and tap into the 4-wire 2,400/4,160 volt primary main. The low-voltage windings are paralleled mside each transformer and then connected in wye and tapped onto a 4-wire, 120/208 volt secondary main.

The construction features of this installation are typical of those found in other three-phase services mounted on one pole, so let's take a close look at them. First of all, notice

Figure 7-5.—Pole mounted three-phase installation

the method of supporting the transformers (fig. 7-5). A double crossarm mounted approximately a feet below the primary main does the trick. Notice that a kicker-arm is installed just below the double crossarm. The kicker-arm provides a support for the bottom of the transformer hanger

and thus maintains the transformer in a rigid vertical position.

Next, let's get a close-up of the lightning arrester connections and construction (fig. 7-6). Although only two arresters are visible, there is another mounted on the other end of the crossarm and connected to the third phase line. The arresters are clamped in a hanger mounted flat against the side of the crossarm that supports the primary main. A wire from the top of each arrester is connected by means of a solderless connector to a phase wire. Notice the drip-loop in the wire. A ground wire leads from the bottom of each arrester and connects to the main giound wire that is stapled to the underside of the crossarm.

>9U'e 7-6.—Lishtning arrester connections.

Wliile you're looking at figure 7-6 you can check the method of connecting a training wire from each primary main down to the primary bushings on the transformers via the primary fuse cutouts. A solderless connector is used to clamp the training wire to the primary main. The wire is then looped dowTi and connected to the top of the primary fuse cutout (fig. 7-7). Pin insulators mounted on the sides of the top crossarm and double crossarm provide a support

Fi3ure 7-7.—Primary fuse cutout connections.

for each training wire. Again, notice the dri]) loop placed in the wire, A wire is trained from the bottom of the fuse cutout to the primary bushing of the transformer.

Figures 7-7 and 7-5 clearly show how the primary fuse cutouts are mounted on the double crossarm by means of a

crossarm hanger. The fuses shown are a combination disconnect switch and fuse. When the hinged cover is pulled | down in the position shown, the training wire circuit leading to the primary bushing is opened. When the hinged cover is closed, the fuse is automatically inserted in the circuit.

TRANSFORMER POLARITY

Standard markings have been adapted for transformer terminals. The terminals for the high voltage, or U, winding are marked Hi, Hj, and so on. The terminals for the low voltage, or X, winding are marked Xx, X2, and so on. These polarity markings indicate the order in which the leads are ' brought out from the windings and also the respective polarities of the primary and secondary leads at any instant. If a transformer winding is marked //i, H^y H3, and H4, the end leads (full winding) are indicated by Hi and H4', the intermediate taps are indicated by H2 and H3. As you fa^e the high-volt'age side of a transformer, the extreme right-hand lead is marked H.

When the Hi and Xi leads are brought out on the same side of the transformer, the polarity is called subtractive. If the Hi and are connected together and a voltage is applied across the Hi and H2 leads, the resulting voltage across the H2 and X2 leads in the series circuit formed by this connection equals the difference of the voltage of the two windings. The voltage of the H winding opposes the voltage of the X winding—hence, the term subtractive polarity.

When the Hi and X2 leads are brought out on the same side of the transformer (as in fig. 7-8), the polarity is called ADDITIVE. If the Hi and X2 leads are connected together and a voltage is applied across the Hi and H2 leads, the resultant voltage across the H2 and Xi leads in the series circuit formed by this connection equals the sum of the voltages of the two windings. The voltage of the H winding aids the voltage of the X winding—hence, the term additive polarity. Polarity

markings do not indicate the internal voltage stress in the windings, but they are useful in making external connections between two or more transformers.

Both the primary and the secondary windings of a transformer may consist of two or more windings connected in series or in parallel to supply several different voltages. If the windings are not connected properly with relation to ^ach other, a virtual short circuit ma\'7d' result when the transformer is put in operation. When two or more transformers of the same polarity are operated in parallel, similar leads must be connected to the same line wires to prevent short-circuit currents from flowing in the local circuits between the transformers.

A schematic drawing of three single-phase transformers with both primary and secondary windings wye-connected is shown in figure 7-8. The connections are for additive

B
UIJMMMM
H2
X2
nmn
Hi
N
[i).().O.QOOOQOQO()(U
H2
PPPT]

nnnp
A» n' B'
Figure 7-8.—Schematic wye-wye connection.
C»

polarity. The Hi leads of each primary winding are brought out as the respective A, B, and C phase leads. These phase leads are connected to the corresponding line wires. The A'2 leads of each secondary winding are brought out as the respective A\ B', and C phase leads. These phase leads are connected to the correspoiuling load terminals. The Xy leads of these windings are connected together to form the

neutral of the wye connection. In a three-phase, four-wire system (which is discussed here), the fourth or neutral wire is connected to this mid point (or neutral). Such a system will serve both single-phase and three-phase equipment. For example, single-phase power at 120 volts may be obtained across any one-phase wire and the neutral, and three-phase power at 208 volts may be obtained across the three-phase wires.

For subtractive polarity, the and H2 leads are connected in the same manner as for additive polarity. However, the Xx leads are brought out to form the respective A', B', and C phase leads, and the leads of the secondary windings are connected together to form the neutral of the wye connection.

THREE-PHASE TRANSFORMER BANK INSTALLATION
(PLATFORM)

When the combined weight of the transformers cannot be safely handled by one pole, the transformers are placed on a platform supported between two poles. Transformers rated 25 kv.-a. through 100 kv.-a. can be installed in this manner. Except for the difference in construction, the general principles of connecting the transformers and the protective equipment requirements hold equally as well for this type of installation as for the pole-mounted type.

The three-phase transformer bank illustrated in figure 7—9 is a typical example of a platform-mounted installation. Both a front and side view are shown and the physical dimensions and electrical requirements are clearly indicated. The platform's foundation is built of two 4- by 8-inch creo-soted timbers mounted between the poles and bolted on each end to a pair of 4- by 8-inch creosoted crossarms. Two by fours, spaced 1 inch apart, serve as the floor of the platform. Of course, the dimensions of the timber given can't always be met by the stock you have on hand. But they should be matched as closely as possible. By the way, it won't be necessary to bolt the transformers to the platform. Their weight will be great enough to keep them in place.

In this particular example, the transformer bank is placed at the end of the primary main run. Therefore, it's necessary to use a pole-to-pole guy and pole-to-ground guy to counteract the pull of the wires. The three-wire, 2,400-volt primary main is brought in from the right and dead-ended on suspension-type insulators on the top crossarm.

Fi3ure 7-9.—Platform-mounted three-phase transformer bank.

Three lightning arresters clamped to the top crossarm are connected to the primary main and to a ground wire on the side of the pole. Training wires tapped to each primary main drop down and connect through fuse cutouts to a tlu-ee-wire primary bus stretched between the poles. The secondary bus is also mounted between the poles but above the primary bus.

The transformer bank is connected with a delta prinuiry and a delta secondary. A four-wu'e secondary main for light and power is obtained by mid-tapping the secondary of one of the transformers.

THREE-PHASE TRANSFORMER BANK INSTALLATION (GROUND)

The larger single-phase transformers when connected into a three-phase bank must be set directly on the ground. This situation will arise in some substation installations. You may also find it necessary to install transformers on the ground if two three-phase banks are needed to supply power to the same area or building. Aside from the special construction features necessary to mount a transformer on the ground, there are only two major safety requirements which make a ground bank installation different from a pole or platform mounted bank. One, in addition to the normal grounding of neutral wires and lightning arresters, all equipment and steel work must be effectively and permanently grounded. Two, a protective fence must be built around the installation and warning signs prominently displayed.

Let's look at the actual step-by-step construction of a ground-installed three-phase bank. But before you start here is an over-all picture of the installation. A large quon-set hut (fig. 7-10) requires power for (1) electrical equipment rated at 120/208 volts, single-phase and three-phase, (2) electrical equipment rated at 240 volts, three-phase, and (3) electrical equipment rated at 2,400 volts, three-phase. The power for this equipment is brought over to the building on an overhead 2,400-volt, three-phase delta primary main (fig. 7-10). The 2,400-volt three-phase power is then brought down from the overhead primary main and sent into a transformer bank installed beside the building. At the transformer bank, the 2,400-volt power takes two paths. One route goes directly into the building via a set of fused cutouts. This arrangement provides power' for the electrical equipment in the building rated at 2,400-volts three-phase. The other route takes the 2,400-volt power through another set of fused cutouts and distributes it on a primary three-wire bus. The primary bus serves two sets of transformer banks. One bank, consisting of three 25 kv.-a. transformers, connected delta-delta, steps down the 2,400-volt three-phase power to 240-volt three-phase power. This 240-

Fi3ure 7-10.—Overhead primary mains bring power to quonset hut.

volt power is then sent into the buildint? to serve the 240-volt three-phase eh'ctrical e(iiiipinent. The other transformer bank is made up of three 10 kv.-a. transformers. They reduce the 2,400-voil j)o\ver to 120/208 volts by means of a deha-wye arrangement. Tliis seeondarv vohag:e is carried into the building over a four-wire secondary main and is then

distributed to equipment rated at 120 or 208 volts, single-phase or three-phase.

CONSTRUCTION DETAILS

Now for the construction details. The transfer of the 2,400 volts from the overhead line down to the transformer bank on the ground is to be accomplished by a lead-sheathed, three-conductor, No. 1/0, varnished cambric power cable. Protection of the cable is afforded by a four-inch steel conduit that is secured to the side of the pole with pipe straps (fig. 7-11). At the bottom of the pole, the conduit goes underground to the right and emerges at the transformer bank site.

Figure 7-11.—Transformer bank site. 200

The transformer bank site is shown in figure 7-11. It is in tlie initial stage of construclion. The conduits which protect t he wires entering tlie building have been placed in position. AAi excavations have been filled in and the site leveled. The type of power carried by the wires througb each conduit is clearly labeled. Notice that the ends of the conduit have l^eon plugged. This keeps dirt and other material used during construction out of the conduit line.

The next construction step involves the steelworker and builder. You'll be around, of course, to supply any information the\'7d' might need. The steelworkers job will consist of fabricating the steel supporting frame and posts shown in figure 7-12. They will also lay and tic in the reinforcing rod used to strengthen the concrete slab. The l)uiI(hMs will build the wooden forms that hold the concrete until it is set. They will also place the supporting frame and poles in their correct place and pour the concrete.

Fisuie 7-13.—Placing the stool supports.

The steel supports shown in figure 7-12 are fabricated from 3- by 3-inch angle-iron posts. The rectangular frame will support the fused cutouts that are inserted in the primary main coming from the pole. That's why the frame has been centered over the end of the conduit that leads down from the pole. The other two angle-iron posts, with the help of the frame, will act as supports for the primary bus that feeds the transformer banks. The frame and posts are set in holes and after being leveled and plumbed are shored to maintain their position while the concrete is being poured.

After the builders and steelworkers have done their job, the next step is all yours. It consists of preparing a grounding system network which will permit you to ground the equipment after the concrete has been poured. Start by driving a ground rod at least 8 feet into the ground. A closeup of this installation can be seen in figure 7-13. Then, starting from the ground rod, train a ground wire over to the steel work that is to be grounded. This includes'the angle-iron posts, angle-iron frame, and conduits. As shown in figure 7-13, the ground wire network is laid loosely under the reinforcing rods. You can also see how the ground wire is brought up and temporarily secured to each piece of steel work that is to be grounded. After the concrete has been poured, these ends will be available for permanent clamping to the equipment.

After you've completed the grounding system, the builder steps in again this time to pour the concrete. As soon as he says the concrete has set, however, you've got to come back and (1) mount the fused cutouts and the insulators, (2) wire the primary bus and the fused cutouts, (3) install the transformers, (4) connect each transfoi*mers bank for delta and tap into the primary bus, (5) connect the 75 kv.-a. bank for a delta secondaiy and pull its secondary wires into the building through tlie conduit provided for this purpose, (6) connect the 30 kv.-a. bank for a wye secondary and pull its secondary ^vires into the building through the conduit provided for this purpose, (7) pidl a lead-sheathed, three-cond\ictor. No. 1/0 cable into the building and connect it by means of a pnjthead to one set of the fused cutouts, (8) permanently ground I lie

Figure 7-13.—Croundins network.

equipment and steel framework, (9) install the lead-sheathed power cable in the conduit between the overhead primarj^ main and the transformer bank, (10) connect tliis power cable, by means of potheads, to the decnergized primary main and to the other set of fused-disconnects in the transformer bank. When all these steps have been completed, the installation will appear as in figure 7-14.

Part of the completed installation is shown in figure 7-15. Here you can see how the fused cutouts are mounted on the rectangular steel framework. A miniature three-wire bus supported by pin insulators is connected between the cutouts. This bus provides a tapping-in-point for the wires carrying the 2,400-volt primary voltage down from the overhead primarj' main. The bottoms of the fused cutouts on the left are connected to a pot head located directly beneath the cutouts. This pothead provides the interlink between the

2,400-volt, tliree-wire, lead-sljoathcd cable leading into the building and the open connecting wires of the cutouts. The bottoms of the cutouts on the right are connected to the three-wire primary bus that is stretched the length of the transformer bank. Notice how the primary bus is supported on and tied to the insulators.

Figure 7-14.—The completed installation.

Also visible in figure 7-15 is the delta arrangement of the high-voltage terminal of the three 25-kv.-a. transformers. The Hi terminal of the first transformer is tapped into the top wire of the primary bus. The Ih terminal of the same transformer and the II\ terminal of the center transformer are tapped into the middle primary bus wire. The Hi terminal of the center transformer and the Hi terminal of the third transformer are tapped to the bottom primary bus wire. Finally, the Ih terminal of the third transformer is tapped back into the top primary bus wire. What it amounts to, is connecting the high-voltage windings of the three transformers in series to form a closed loop and tapping off at the junctions to the primary bus. The low-voltage terminals of the three 25-kv.-a. transformers are also cod-nected delta. Therefore you'll follow the same general connection procedure as for the high-voltage terminals.

Figure 7-15.—View of 75 kv.-a. bank.

The ,secondary wires are sent into the conduit through a protective service liead.

Figure 7-16 is a view of the other end of the transformer bank installation. Here can be seen the construction details of the three lO-kv.-a. transformer. The high-voltage terminal connections are hidden in the picture, but since

Figure 7-16.—View of 30 Icv.-a. bank.

they are delta-connected their arrangement will be exactly like those of the three 25-kv.-a. transformers. Notice how the low-voltage terminals ai-(» arranged together to obtain a wye secondary. Corresponding terminals of each of the three transformers are connected together and grounded. The remaining three free terminals are connected to the secondary line wires wliich enter the conduit service head. While you're at it, check the groimding connection at the conduit. Notice how ground clamps are used to insure a permanent bond.

TRANSFORMER MAINTENANCE AND REPAIR

The largest part of your transformer maintenance and repair wQl be concerned with (1) the inspection and replace-ment of defective bushings, (2) the inspection, and-replace-ment of insulating oil, and (3) the reconditioning of leads.

Transformer installations should be inspected at regular periods. In addition, you should instruct your crew to keep an eye open for any indications of transformer failures during their daily work routine.

BUSHING REPAIR

Defective bushings will be your number one transformer trouble. The comparatively brittle porcelain covering is always exposed to mechanical damage. Once the glazed surface has chipped or cracked the breakdown voltage rating decreases and arcing results. A climbing or ground inspection will usually reveal the defective bushing. However, arcing at the terminals should be investigated as a possible indication of a defective bushing.

The replacement of a bushing is not advisable while the transformer is mounted on the pole. Your working space is limited, thus increasing the possibility of an accident. If possible, replace the damaged transformer and take it back to the repair shop.

Replacing a distribution transformer bushing involves six general steps:

1. Removing the internal winding leads from the defective bushing.
2. Releasing the clamping device that holds the defective bushing in place.
3. Removing the defective bushing.
4. Inserting a new bushing.
5. Clamping the new bushing m place.
6. Connecting the internal leads to the new bushing.

Figure 7-17 shows yoii the steps in removing a tank-wall bushing. In step one, the internal leads are disconnected from the defective bushing terminal. If a handhole is available, of course, you won't have to remove the tank cover. Step two consists of releasing the clamping device on the bushing. A hook spanner wTench is the tool you'll use for

Figure 7-17.—Removing a bushing.

this particular job. In some transformers, the clamping device consists of a circular spring and plate affair that applies pressure to the bushing from inside the tank. Step three shows the defective bushing being removed from the tank. Notice that cork gasket. It should be examined and replaced if found to be in bad condition. The new bushing is replaced in opposite order to that used in removing the defective bushing. Therefore, when working on a bushing of an unfamiliar type go slow. Then you'll remember all the details of the removing process, and can apply them, in reverse, to the replacing process.

TRANSFORMER OIL

An important function of the oil used in transformers is to provide insulation between the windings and thus prevent high-voltage arc-overs. The dielectric strength of the oil

therefore, must be maintained at a high level if equipment failures or "burnouts" arc to be avoided. The dielectric strength of the oil will depend on the oil's relative pureness. Impurities, such as water, solids, and sludge lower the dielectric strength of the oil if present in appreciable quantities. These impurities form after the transformer has been in service for some time.

Periodic inspections will help save a lot of transformers from the junkpile. The larger transformers are equipped with temperature and Hquid level gages that provide a visual check of the oils' condition. The condition of the oil in the smaller transformers can be determined by feeling the wall of the transformer tank. If it is excessively warm, the oil, in all probability, contains large amounts of water.

If you suspect that the transtormer oil is impure remove the transformer from the line and return it to the repair shop for conditioning.

At the repair shop, a test is made of the dielectric strength of the transformer oil. In general, this involves the determination of the high-voltage breakdown point of the oil.

A portable oil tester, figure 7-18 is used for this test. To test a sample of oU proceed as follows:

1. Set the electrode gap in the testing receptacle (cup) to 0.100 inch. A gage is provided for this purpose.
2. Fill the cup with a sample of the oil to be tested. Be sure t« thoroughly rinse the cup

with the oil.

3. Plac« the cup in the bushiags provided for it in the tester (step r, fig. 7-19).

4. Lower the guard over the cup.

5. Turn the potentiometer knob to the left as far as possible and set the circuit breaker by raising the breaker handle on the left side of the oil tester's panel.

6. Hold the spring switch closed and increase the voltage (as read by meter) to 25 kv. by turning potentiometer to the right (step 2, fig. 7-19).

If the dielectric strength of the oil is not up to par, an are will jump the electrode gap before the 25 kv. mark is reached. When that happens, the circuit breaker automatically goes into action and opens the testing circuit.

Transformer oil that does not meet the testing specifications must be removed and replaced with fresh oil. For small single-phase transformers, the best method of removal is to siphon the major part of the oil from the transformer tank and then tip the tank over to drain out the remainder.

After the tank has been drained, it's a good idea to clean the transformer. A more thorough job can be obtained if the transformer is removed from the tank. The coils, core, and terminal board are then washed down with fresh oil under pressure. Small units may be cleaned with a soft brush saturated with fresh oil. The clean transformer can then be replaced in the tank and new oil added to the correct level. As a word of advice—always use metal pipes never rubber hose. Rubber reacts with the oil and a harmful compound is developed.

If an oil tester is not available, a simple, though not as reliable, test can still be made to detect the presence of moisture in the oil. A red-hot nail or poker thrust into the oil will cause the oil to "crackle" if water is present.

Oil that has a low dioloctric strength can be reconditioned if the repair shop is equipped with a blotter-tj^pe filter press. One such type of press is sho\vn in figure 7-20. Its main parts consist of (1) a motor (2) a pump and (3) a series of filter paper. The pump, driven by the motor, forces the dirty and wet oil under pressure through the blotter type paper filtere. Dirt is screened out and water is absorbed by the filter paper. The clean oil is directed, by means ot pipes and valves, to a storage tank. A pressure gage is used as an indication of the filter paper condition.

COHPRISSION SCIEK

Figure 7-20.—Blotter-type filter press.

When the fdter paper becomes clogged with dirt the back pressure uicreases. Therefore, when the pressure gage reads more than 70 pounds, the filter paper should be renewed. All pipe connections and specific operational procedures will he found in the manufacturer's manual accompanying the filter press.

RECONDITIONING LEADS

External transformer leads require no cutting when they have to be disconnected from transformers equipped with

Figure 7-21.—Splicing fransfortner leads. 213

clamp terminal bushings. The external leads are simply disconnected at the bushing terminals and the transformer is then ready to be installed at another location. However, in the older types of transformers the internal winding leiwis are brought directly through the bushings and thus serve also as the connecting leads. Each time that it becomes necessary to remove the transformer, the external leads must be cut. Eventually, the leads will become too short to work with. Then you'll have to replace them by splicing new lead extensions onto the internal winding'leads.

The major steps in splicing a transformer lead are shown in figure 7-21. Step one shows the extension lead being secured to the winding lead by the use of tinned copper seizing. Seizing is employed in this particular case because the extension lead is a round conductor and the winding lead a flat conductor. If both conductors were round, a crimping cormector could be used. About ten turns of the seizing ribbon are taken and the end tucked imder the last turn.

Step two of figure 7-21 illustrates the method of applying i solder to the two spliced leads. The "wiping" solder is poured from one ladle to another until a fairly heavy coating covers the splice. Of course to get any "tinning" action at j all it will be necessary to apply a thin film of flux paste to the bare leads before the pouring begins. Notice that masking tape has been wrapped over the insulation of each conductor · This prevents charring of the insulation by the hot solder. ! Also notice the cloth that protects the transformer coils. \

The splice is usually wiped. This step consists of holding | a wiping cloth underneath the splice and pouring a little i solder over the splice. The wiping cloth is used to mold the plastic solder into a smooth even covering. You should, under no circumstances, tape the splice. A break in the continuity of the insulation is necessary to prevent the transformer oil from being drawn out of the case by a "wick" action.

QUIZ

1. You are given the job of installing a single-phase transformer in a certain area on an advanced base; there are no drawings available. What is the first step you should take?

2. Where should the transformer be located?

3. How would you install a 200-kv.-a. transformer?

4. What size fuse would you use in the primary cutout of a lOO-kv.-a. transformer that operates from a 4,800-volt primary?

5. How are platform mounted transformers secured to the platform?

6. What major safety requirements make a ground bank installation dififerent from a pole or platform mounted bank?

7. What is the chief transformer trouble?

8. What steps are involved in the replacement of the bushing?

9. What purpose does the oil in the transformer serve?

CHAPTER 8

ADVANCED BASE PORTABLE ELECTRIC POWER

PLANTS

YOUR JOB

The purpose or objective of an advanced base is easy enough to define. It is to support the advancing iVrmed F'orces. The size of an advanced base is a little harder to pin down. The reason becomes obvious when you stop to consider that an advanced base might be just a small seaplane refueling outpost or an installation as large as Pearl Harbor. How does this affect you? Well, suppose you were stationed at that small base. Would you know how to supervise the installation, operation, maintenance, and repair of portable generating equipment? Or if you were assigned to the larger more permanent base, woidd you be capable of assuming the duties associated with the operation of a central electric power station? These responsibilities are yours as a CE, CEG, or CEP in the rates of first class and chief.

A REVIEW

Chapters 5 and 7 of Construction BJlectrician^s Mate 3 and 2, NavPers 10636-A introduced you to advanced base generating equipment. If that introduction is just a passing memory then by all means renew your acquaintance with those chapters. However, if all you need is a summation of the important points contained in those chapters check the following items:

1. The three basic components of an advanced base electric power plant are (1) an internal combustion prime MOVER that develops mechanical power from heat power, (2) an alternator that converts the mechanical power of the prime mover into electrical power, and (3) a SWITCHBOARD that contains meters, switches, and other accessories necessary for the control of both alternator and prime mover.

2. Advanced base electric power plants are typed as either PORTABLE units Or PERMANENT UnitS.

3. The term "packaged power" aptly describes the portable plant. Alternator, prime mover, and switchboard make up a self-contained unit that is usually enclosed in a sheet metal housing with removable side panels. The complete plant when moimted on steel skids is capable of being moved from one location to another with a minimum of trouble.

4. The permanent electric power plant is designed for protected installations in central generating stations. The switchboard is separated physically from the generating unit but connected electrically by means of wires placed in conduit.

6. The internal combustion engine of the piime mover may be of the gasoline or Diesel type. The gasoline type is designed for advanced base electric plants of low power output (1.5 kw. to 15 kw.). The Diesel prime mover drives the larger generating units (30 kw. to 700 kw.).

6. Gasoline engines employ spark plugs to ignite the combustible air-gasoline mixture in the cylinders. The
smaller gasoline-driven electric power plants are hand-cranked. The larger engines require an electric starting motor similar to the type used in automobiles.

7. Diesel engines have no ignition system. Combustion in the cylinders is developed by the high heat of compression. This feature makes Diesel engines hard to start. The portable plants are started by three different methods depending on the manufacturer of the engine. One method employs an auxiliary gasoline engine, another method converts the Diesel ei^ine into a temporary gasoline engine, and a third method uses an electric starting motor. The large permanent plants use compressed air for starting purposes.

8. The frequency of the alternator's voltage output depends on the speed of the prime mover. The speed of the prime mover is controlled by the amoimt of tud admitted to the cylinders.

9. The strength of the alternator's voltage output depends on the strength of the excitation voltage supplied by the exciter. The excitation voltage output of the exciter is controlled by the amount of resistance placed in the exciter's field coils.

PORTABLE ELEQRIC PLANTS

When an advanced base is first established, electric power is needed in double-quick time. The point is, you won't have time to build a central generating station. Rather, you will spot a portable plant at each important location requiring power. What type do you use where? Well, just let your common sense be your guide. For example, the electrical load at the headquarters building will consist of lights and small single-phase motors. There's no need to waste a three-phase Diesel-driven plant at this location when a small gasoline-driven job will do as well. That Diesel-driven plant will come in handy at the sawmill or wherever large amounts of single-phase and three-phase power are required.

Naturally, the power and voltage requirements of the load will determine the size of the plant to be used. Electrical equipment rated at 110/120 volts, single-phase, with a combined power demand of less than 1..5 kw. can be handled by a o:asoline-driven plant with a 120-volt two-wire output and rated at 1.6 kw. (fig. 8-1).

Figure 8-1.—Single-phase gasoline-driven generator set.

Other types of generatoi's used at advanced bases are as follows:

1. 10-kw, 120 volts, d-c, two-wire, gasoline-engine powered

2. 5 and 10 kw, 120/208 volts, three-phase, four-wire, 60-cycle Diesel-engine powered

3. 15, 30, 60, and 100 kw, 120/208 volts, three-phase, four-wire, 60-c3'cle, Diesel-engine powered

The 5 and 10 kw a c generators are provided with means for reconnecting the generator leatis to produce 120 volt,

one-phase, two-wire or 120/240 volt, one-phase, three-wire, 60-cyclc current. The 15, 30, 60, and 100 kw generators are provided with means for reconnecting the generator leads to produce 240/416 volts, three-phase, four-wire, 60-cycle current. All of these generators are skid mounted.

In the past the Seabees have used generators rated at different voltages from the ones just described. Examples are generators designed to produce 220/440 volts, three-phase rated output. The principles of installation sucli as determining cable size and generator connections are the same regardless of voltage rating. Conveision formulas and tables based on the National Electric Code will give you the proper information.

EMERGENCY INSTALLATION FACTS

Figure 8-2 should convince you that there's nothing fancy about an advanced base emergency electric power plant installation. The important thing to remember is set the INSTALLATION UP IN THE SHORTEST TIME POSSIBLE. Keep

Figure 8-2.—Diesel power plant on airport runway.

in miDd, however, that the plant is also expected to give good service for an indefinite length of time. Therefore, if you find it necessary to improvise, make it last. For example, if you substitute bottles for insulators make sure the bottles are properly secured. And if you're out of regular wire and have to use communication field wire make sure that you parallel enough of the field wire to safely carry the load current.

In an emergency installation, pick a site that is as close to the load as possible. Although a concrete deck is desirable, there may not be enough time for the builder to construct one for you. So do the next best thing and have your crew set up a wooden platform on which the skid-mounted plant may be placed. In either case, make certain the plant is leveled. And try to take advantage of prevailing winds (if any) when you place the plant so as to obtain the most satisfactory cooling of the engine.

INSPEQING AND SERVICING THE PRIME MOVER

If you are installing a new plant, your crew must do some preliminary work before placing the plant in operation. First they should make a visual over-all inspection of the plant. Have them look for broken or loose electrical and hose connections and loose bolts and capscrews. If any wiring connection is suspected of being faulty, check it against the wiring diagrams in the instruction book furnished with the power plant. Any trouble that is found should be remedied immediately.

Servicing the prime mover is the next step in the process of placing the plant in operation. Fill the crankcase with lubricating oil to the "full" mark on the bayonet gage, or dip stick. A lubrication chart in the instruction book furnished with each plant will recommend the proper grade of oil to use in accordance with the operating temperature. While the crankcase is being serviced, have another part of your crew fill the radiator with water. If the plant is to be operated in freezing temperatures use an antifreeze solution in the proportions recommended in the plant's instruction book.

If you have enough of a crew to work with, you can probably spare a few men to fill the engine's fuel tank while the other servicing is going on. Make sure that clean fuel is used—strain it if necessary. Some of the Diesel units will have a separate main fuel tank that supplies fuel to a smaller tank mounted on the prime mover. Two lengths of copper tubing are provided with the unit (one for the suction line the other for the return line) for connection between the main fuel tank and the engine. Figure 8-2 shows this setup.

Some of the prime movers of advanced base electric power plants are started by starting units which obtain their power from batteries. If the plant that you are installing is equipped with a battery (or batteries) your crew has another servicing job to do. Batteries are usually shipped without the electrolyte in them, but with the plates in a dry-chai^ed condition. Thus, it is necessary to fill the battery with electrolyte. Usually, the electrolyte is shipped right with the plant and is of the correct specific gravity for that type of battery furnished for the plant. However, if you must prepare your own electrolyte use the mixing chart in table 2.

Tabic S.—Elcdroiyt* mixing chart

The specific gravity of the electrolyte depends on the type of battery furnished with the plant. Use the specific gravity

value as recommended by the plant's instruction book. Generally, you'll find that batteries with wood separators require an electrolyte with a filling gravity of 1.255 and batteries with rubber separators a filling gravity of 1.200.

After the electrolyte has been prepared, follow the plant's instruction book with regard to the recommended filling procedure. In the event these are not available use the following general procedure:

1. Add electrolyte to each of the battery's cells until the level of the electrolyte is visible in the filler neck or at least % inch above the separators. The temperature of the electrolyte when placed in the cells should be between 60° F. and 90° F. It should never EXCEED 90° F. (32° C).

2. Chemical reaction will cause the battery to become heated. It may be cooled artificially or allowed to stand at least 1 hour before placing it in service.

3. You will probably notice at the end of the cooling period that the level of the electrolyte has dropped. This is due to the electrolyte soaking into the plates and separators. Before placing the battery in service restore the electrolyte to its proper level.

4. Any electrolyte spilled on the battery should be removed with a cloth dampened with a solution of bicarbonate of soda and water.

Although the battery can be placed in service 1 hour after filling it with electrolyte, it is a procedure that you should consider only in an emergency. If at all possible, the battery should be given an initial light chaise. In the event that the charging procedure is not covered in the plant's instruction book, use the following general method:

1. Charge the battery at a low rate (about 5 amperes) until the voltage and specific gravity, corrected to 80° F., have remained constant for at least 5 hours.

2. If the temperature of the electrolyte reaches 125° F. reduce the charging rate or stop the charge until the battery cools.

3. During charging replenish any water lost by evaporation.

After the battery has been chained it is connected into the starting system of the prime mover as indicated by the wiring diagrams accompanying the plant.

INSPEQING AND SERVICING THE ALTERNATOR

Just as important as the preparation of the prime mover for operation is the inspection and servicing of the alternator. Generally, you should—

1. Check all electrical connections with the plant's connection diagrams.

2. See that the connections are tight.

3. See that the collector rings are clean and have a pohshed surface.

4. Check collector brushes to make sure they have no tendency to stick in the brush holdero, that they are properly located, and that the pigtails will not interfere with the brush rigging.

5. Check the collector brush pressure to see if it agrees with the figure recommended in the plant's instruction book. In the event that the brush pressure information is not available, use a brush pressure of approximately 2 p. s. i. of brush area.

6. Check the exciter in the same manner as the alternator.

SOMETHING TO THINK ABOUT

The inspection and servicing procedures that you've just read are rather general. In most cases they can be applied to any electric power plant that you install. You realize, of course, that there are other special installation details which are pertinent only to the particular plant you happen to be working on. Because of the many different types of power plants these special details cannot be discussed in this chapter. Your source of information for these details is the manufacturer's instruction book that accompanies each power plant. Save these instruction books! Additional copies are seldom available.

Keep in mind that you're the one that's going to have to

answer for any failure of the plant due to improper servicing and operation by your crew. It's those small precautions which are usually overlooked that cause most plant failures. Have you made sure that the water used to fill the radiator is clean and soft? If the plant is operating in freezing temperatures, and you were unable to secure antifreeze, are you checking to see that your crew drains the water from the cooling system at the end of each day's run? Have you provided protection for the fuel in the ^uel tank and the windings in the generator? The fuel tank cover and the generator cover in figure 8-2 may be makeshift but they help to reduce plant failures. Are you enforcing safety rules? Details such as keeping the funnel in contact with the fuel tank when filling it with gasoline might save a life and ensure a full working crew.

CONNECTING THE PLANT TO THE LOAD

While the electric power plant is being installed and serviced, a part of your crew can be assigned to the job of connecting it to the load. EssentiaUy this consists of running wire or cable between the plant and the load. At the load end, the wire or cable is connected to the equipment or to the interior wiring system of the building, if that is where the equipment is housed. At the power plant end, the wire or cable is connected either to the output terminals of a main circuit breaker or a branch circuit breaker.

Running the wire and making the connections is purely a mechanical process. Your crew will take care of that. But it wiU be up to you to—

1. Decide whether the wire or cable will be buried or carried overhead on poles.

2. Determine the correct size of wire or cable to use.

3. Check the generator lead connections of the plant to see that they are arranged for the proper voltage output.

The information contained in the following sections will help you in these tasks.

INSTALLING THE LOAD CABLE

In an emergency installation, time is the important factor. So let that be your guide when you have to decide whether the wire or cable will be buried or carried overhead. It is | certainly a lot simpler to just dig a trench and lay the cable in it than it is to erect poles and string wire on crossarms. Of course, the material you have on hand will help you make your decision. Weatherproof wire, type WP, with just a triple-cotton braid insulation, wouldn't last very long if buried directly in the ground. In the event that that is the only type of wire available, you will have to place it overhead. But if rubber insulated, rubber jacketed,, underground cable is available (in the proper size), you can bury it directly in the ground with a minimum of trouble.

When conditions dictate an overhead line be sure your ' men construct it properly. Nine times out of ten, regular wooden poles will not be available. Then you'll have to improvise with whatever is handy. Four-by-fours will do the trick if they are long enough to be set rigidly in the ground and stiU provide safe clearance for the wires. In a real pinch, trees may be the answer to your line-support problem.

Direct burial of cables calls for a few simple precautions to ensure uninterrupted service. They are—

1. Dig the trench at least 18 inches deep to prevent disturbance of cable by subsequent surface digging (fig. | 8-3). i

2. Lay the cable over a sand cushion (fig. 8-3). If this | is impractical, loosen the trench base so it is cleared of rocks and stones.

3. Uniformly separate the cables for further mechanieal and electrical protection. About 6 inches between centers will be sufficient (fig. 8-3).

4. After laying the cable and before backfilling, cover it with earth free from stones, rocks, etc. This will prevent the cable from being damaged in the event the surrounding earth is disturbed by flooding or frost-heaving.

MHO (USNION

Fisure 8-3.—Direct burial of cabi*

DETERMINING CABLE SIZE

There are two things that might happen if you use the wrong size conchictor in the load cable. P'irst, if the conductor is too small to carry the current demanded by the load, it will melt and cause a break in the circuit. A fire hazard is also involved. Second, even though the conductor is large enough to safely carry the load current, its length might result in a lumped resistance that produces an excessive voltage drop. An excessive voltage drop results in a reduced voltage at the load end that is incapable of operating the equipment. In this respect, it might be well to point out that the National Electric Code calls for a voltage drop not in excess of 3 percent for power loads and 1 percent for lighting loads or combined power and lighting loads. For example, if a power plant generates 440 volts for power equipment, the voltage drop in the line should not exceed 13.2 volts (440X0.03).

To determine wire size, then, you must consider (1) the load current to be carried by the cable, and (2) the distance between the load and the power plant. Generally, the following steps are involved:

to

X
■<*< X
X
g
>>
u u X
•—<
X
n >
X
o X
m ■«->
﹀
08
eo
X
o
>
o o o
X
X
o
>
CO X
u
X
CO
-t->>
X
o >
X si
>
o
>
X X
CO
"o
>
X
CO
a
X
> X
OQ
a
O

a

o o o

eo

X

OQ ■<->

o >

a

OB

"o

>

X

CO

0 <

5

eS O

51 03

u t! V

X

eo , •

E X

X ^ as

a

c

CO

CO -<-:>

08

is

•I- -2 > ° o

a S <

a; .

X ^

00 ^

e

X

1 ^

s

o

e

o

^ c

c ^

lU o

a. =s

S

<

.s

S3
O
c
2
c
o c
c
is o c
.23
08 I
> S
00 O)
Ʋ
a
a
CO
C3 f
o
t4
I (2
3
a -«->
3
o a
£
o
s
a
■532
^"3
SI
a> o
M to O
« -4 c>i
o
CO to t-
C4 t^
S3^
228 855? SSS
pi V o
h» »- ^ to O
to e4 CO
lO to
o o
00
t- o»

C4
3^8
;s 8
8 $ S
S S S 2 S
'-' ^ CS (N CO
V2
5
t» to V
ei CO
o e< «o 00 eq
2 8 g
CO CO
ei eo
e) CO CO ^ to 00
n x> ei oj go
M5 us
4< CO
lO —
00 ej to
8 S S
\'7b2 2?
S •? S
<0 O
«
t~ eo to
CO ^
2 e
s
e a
n
^ e«
CO us 1^
2 2 8
8 $
S S i!^
8 8 S?

Step 1. Compute the total current demand with the aid of table 3 and table 4.

Step 2. Use table 5 tafind the size cable capable of carrying this total current.

Step 3. Use table 6 to determine what the lumped resistance of the cable will be when stretched between the power plant and the load.

Step 4. Compute the voltage drop in the cable with the information obtained in step 3.

Step 5. Compare this voltage drop to the maximum figure aUowed by the National Electric Code.

Perhaps an example will help. A rough sketch of a typical emei^ency installation is shown in figure 8-4. A 30-kw. Diesel-driven plant is supplying on-the-spot power to electrical equipment located 50 feet away. The voltage output of the plant is 127/220 volts and is to be carried over to the equipment via four single-conductor rubber-jacketed cables.

The first step is to compute the total load current. This will be the sum of the currents required by the individual pieces of electrical equipment. In this particular example you find the following equipment:

OUOKHMUT

Figure 8-4.—Determinins wire size.

12 100-watt, 120-volt lamps.

1 lO-horsepower, 220-volt, S-phase, slip-ring inductioD motor.

Twelve 100-watt lamps require a total power input of 1,200 watts or 1.2 kw. (12X100 watts). Table 3 indicates that to find amperes when kilowatts are known for single phase a c circuits use the formula:

/=— 1^ kw.X 1,000 p, lighting circuits is 1).

volts X power factor » » /

Therefore, by substitution 7=1.2X1,000

120X1

= 10 amperes for the lamps.

That takes care of the lighting load. Now figure the current required by the motor. Table 4 shows that a 10-horsepower, 220-volt, 3-phase, induction motor has a full-load current of 28 amperes.

Summing up the individual current demands of the equipment you have—

12 100-watt lamps . .. 10 amps

1 lO-horsepower 220-volt three-phase slip-ring induction

motor _ 28 amps

Total load current = 38 amps

Your second step is to choose the cable capable of carrying 38 amperes. Table 5 wUl help you. It lists the allowable current-carrying capacities of conductors of different sizes with different insulation coverings. You are installing undergroimd cable that has a moisture-resistant rubber covering, type RW. Therefore you will use column B of the table. Follow column B down until you reach an ampere value that is greater than 38 amperes. In this case you find a value of 40 amperes. However, since 38 amperes is so close to 40 amperes you should choose the next ampere value (55) to allow for possible additions to the load. Column A indicates that conductor size AWG No. 6 is capable of carrying 55 amperes.

Tobit 5.—Allowable cwrcnt-carrying capadtitt of conductors in Qinpctts

Not More Than Three Conductors in Raceway or Cable (Based on Room Temperature at 30° C. (86** F.))

Tobic 5.—Allowable cuircnt carrying capaclHct of conductors in amperes—

Continued

Not More Than Three Conductors in Raceway or Cable (Based on Room Temperature at 30° C. (86° F.))

From Table 1.—1946 National Electrical Code

Tobl« 6.—^ysical proptftict of condvctofs

Now (step three) you must find the amount of resistance that the load current will meet if you used No. 6 cable. Table 6 lists the physical properties of stranded conductors. It shows that a No. 6 tinned copper conductor possesses 0.427 ohms of resistance in every 1,000 feet. The cable is to run a distance of 50 feet between the plant and the load. However, since the current must travel both ways it will actually pass through 100 feet of cable. Therefore, the total amount of resistance it will meet will be—

„ 0.427X100 ^^.o7 u R= =0.0427 ohms.
1,000

Step four is next. In this step you are to find the voltage drop in the cable. All it takes is a simple application of Ohm's law.

E (voltage drop)=/ (load current) XjB (cable resistance) or £=38X0.0427=1.62 volts

Now (step five) you compare the actual voltage drop that wtU occur in the cable with the value allowed by the National Electric Code. The voltage drop in the cable (step 4) is 1.62 volts. The maximum voltage drop allowed by the code for a combined lighting and power circuit is 1 percent of the source voltage. In this case the power plant generates 220 volts. Therefore the voltage drop should not exceed 2,2 volts (220X0.01). You can see that the actual voltage drop (1.62 volts) is well within this limit, so you can use a AWG No. 6 cable and meet all requirements. If the actual voltage drop had been greater than the allowable value it would be necessary to use the next larger size cable.

GENERATOR CONNECTIONS

When you install a power plant that has a dual-voltage alternator unit, you must make certain that the armature coil leads are properly connected to produce the voltage required by the equipment. Take a look at figure 8-5. It shows an alternator unit that has been disconnected and re-

r

Figure 8-5.—Alternator coil leads.

moved from a tliree-phase Diesol driven power plant. Plainly visible are the stationary armature coils and core (stator) mounted in the main frame of the generator. The a-c voltages generated in the coils are brought through an opening in the pedestal of the frame by means of 10 coil leads (fig. 8-5). Each coil lead is identified by a number stamped on a metal band.

Wlij^ 10 coil leads? Figure 8-6 will help clear up the picture. It's a diagram of the stator coils in their electrical position on the stator core. Notice that there are six sets of coils (A, B, C, D, E, F) arranged in a wye or star pattern. Each leg of the wye is 120 electrical degrees apart and contains two sets of coils. The end of each coil set is connected to a coil lead. The number on each coil end refers to the number stamped on the metal band attached to the corre-

OT7
on
OTS O 11
OW
OW
o w on
EXTBNAi cai laos M noma ir a numib
STAHra) ON A MHAl UNO Figure 8-6.—Sfator coil diosram.

spending coil lead (fig. 8-5).. Count them and you'll find 10 numbers.

In 220/440-volt dual-voltage, three-phase generators the voltage generated in each set of coils, when the prime mover is operating at rated speed, is 127 volts. Thus, by connecting the external coil leads together in different combinations, you can change the external voltage output of the generator. The chart in table 7 gives the exact data needed to make these connections for both three-phase and single-phase voltage outputs.

Let's take a specific example. Suppose it becomes necessary to obtain 220-volts single-phase from a three-phase, dual-voltage generator. Checking table 7 you find that coil leads T2 and T8 are to be connected together and then in turn, connected, through the main circuit breaker, to one of the two load cables. Similarly coil leads T3 and T9 are to be secured and connected, through the main circuit breaker, to the other load cable. Terminal lugs on the end of each

237

1 I

T3 .

00

II E-

9

it< lO «o

o

2 S

s

on

i

CO

E-

E-

T3

e OS

ec

E-i

r"

3

w

oo

e o

i9

g

c c o

a

E^

00 E--c S

M

E-

E^

c

S

00

H

c OS

CO Eh
eo Eh
3
O T
O
CO
9)
a
o
.s
o
M N N — M
o ei
a I
G
4)
11
> ?
O (N
238

coil lead provide an easy means of bolting them together. Just be sure that the connections are thoroughly insulated with a wrapping of rubber tape followed by a wrapping of friction tape. Table 7 indicates that leads Tl, T4, T7, and TO are not used for single-phase service. Therefore, to finish the job, you should individually insulate each of these four leads with rubber and friction tape. Figure 8-7 (a) shows you how the connections appear on the stator coil dia^am. Figure 8-7(b) shows the appearance of the actual connections.

PROCEED WITH CAUTION

When you reconnect generator leads to meet certain load conditions, you change not onJy the voltage output of the generator but also its current rating. These voltage and current changes will have an effect on the operational characteristics of the switchboard controls and instruments. Thus, before the generator can be put into operation, certain internal alterations must be affected.

For example, suppose that it becomes necessary to recoii-nect a 75-kw., 3-phase, 220-volt generator so that it will have an output voltage of 440 volts. In its original connection (220-volt output), the generator was capable of carrying a full-load current of 264 amperes. Changing over to 440 volts, however, reduces its full-load current to 132 amperes. The first thing you'll want to check, then, is the fuse rating in the main circuit breaker. You'll probably find that for its original connection—220 volts, 264 amperes—the generator was protected by a 275-ampere trip element (fuse). To protect the generator for its new current rating (132 amperes), it will be necessary to replace the 275-ampere fuse with one having a 150-ampere rating.

Another thing to consider when reconnecting a generator is the switchboard ammeter instrument that records the current output of the unit. Since it is impractical to use an ammeter capable of carrying the full-load current, instrument transformers are employed. The instrument transformer reduces the current value to one that may be safely measured -

by the! ammeter. Usually, the instrument transformer is designed to reduce the full-load current to a secondary current of 5 amperes. In turn, the anameter, which is connected to the transformer's secondary, is designed to produce a full-scale deflection with an input of 5 amperes. Since the ratio between the load current and the current in the instrument transformer secondary is practically constant, the ammeter scale can be cahbrated to read a true value of load current. Now, in the example being used, the generator has a full-load current of 264 amperes when it is connected for a 220-volt output. That means that the instrument transformer will induce 5 amperes of current in its full secondary winding when 264 amperes flows in the hne and, as a result, the anmieter will have a full-scale deflection. However, on reconnecting the generator for 440 volts, the full-load current is reduced to 132 amperes. Unless you alter the instrument transformer connections, full-load current will produce only a half-scale deflection of the ammeter and a resultant false reading. In this particular case, the alteration consists of moving the instrument transformer winding connection from the full-winding position to the mid-tap position." Most switchboard ammeters have an additional calibrated scale which makes It unnecessary to apply a correction constant when the current range is changed.

Another alteration that you may have to make in the generator's switchboard wiring concerns the voltage regulator. Essentially, the purpose of this device is to keep the generated voltage within certain limits regardless of changing load conditions. Although there are many types of voltage regulators they all respond to variations in rated line VOLTAGE. However, the hne voltage is not apphed directly to the regulator unit. A potential transformer is employed to reduce the hne voltage to a standard value that is safe to use on the regulator. The ratio of primary voltage to secondary voltage must be kept constant if the regulator is to function properly within the limits of the generator's rated voltage. Therefore, when the output voltage of the generator is changed, you must also change the tap connections on the primary of the potential transformer.

Probably tlie big question in your mind is, "Do these alter- ! ations require that you actually dive into the maze of wiring behind the generator's switchboard and move leads from one terminal to another?" The answer is yes—but only iii a few cases. The majority of manufacturers have made provisions that simplify the change-over from one voltage j output to another. Typical of these change-over provisions is the one employed in the Caterpillar Diesel generating set (fig. 8-8). EJssentially, it consists of a 6-bladed, double-throw VOLTAGE SELECTOR SWITCH mount«d on a terminal board. As shown in the end view of the generating set (fig. 8-8), the terminal board is mounted in the lower right-hand comer of the frame. The three vertical rows of the switch contacts extend through the terminal board and are connected per-

Figure 8-8.—Voltage selector switch on Caterpillar Diesel senerotins Hi.

manently, by means of wires, to the generator's stator coil leads; the current and potential transformer windings; the instruments and controls on the switchboard; and the circuit breaker. The arrangement of the wires are such that when the switch is thrown to the left the stator coil leads are automatically connected to produce an output voltage of 220 volts and the current and potential transformers windings are tapped at the correct point for proper operation of the switchboard's instruments and controls. When the selector switch is thrown to the right, the wiring arrangement automatically connects the stator coil leads for a 440-volt output and taps the instrument transformer windings for correct outputs in accordance with the generator's new voltage and current range.

Another type of change-over device in advanced base generators is shown in figure 8-9. It is a change-over block that is mounted on the rear of the generator's switchboard (control cabinet). The terminal studs that protrude from the change-over block serve two purposes—they provide a disconnect point between the load cables and the generator coil leads and they present a convenient means of altering the operating characteristics of the generator's components without changing the positions of the wires. Notice that each of the 10 generator stator coil leads are attached to a correspondingly nimibered terminal stud. Rearrangement of the coil leads becomes a simple process of interconnecting the terminal studs in a definite pattern by the use of changeover LINKS (fig. 8-9). When in position on the coil lead terminal studs, the change-over links also contact other studs that are connected to such components as the current transformers, and the potential transformers. Thus, alteration of these component's output is also affected automatically. Connection diagrams on a nameplate attached to the changeover block provide the necessary information as to the position of the change-over links for specific voltage outputs.

Again, remember that you are responsible for the proper operation of the generating unit. Therefore proceed with caution on any reconnection job. Study the wiring diagrams of the plant and follow the manufacturer's instructions to

LOAD CABLES
Figure 8-9.—Chanse-over block of o General Motors generating set.
the letter. Before you start tl)e plant up and throw the circuit breaker, do a double-check on all connections.

THE BUS BAR
There are a number of reasons why you may find it necessary to use a bus bar when you set up a portable generating station. For one thing, you may not be able to acquire a generating plant that has sufficient capacity to meet the total power demand of the electrical load. It will be necessary, then, to use two or more generating units and collect their paralleled outputs at one central point (the bus bar). Or, you

may discover that the electrical equipment in the advanced base is scattered in such a manner as to require the use of feeder (branch) lines that can be controlled from a central source. Again, the bus bar is the answer.

A typical advanced base generating station using a bus bar is shown in figure 8-10. The two generators are leveled on a concrete apron that is sloped for drainage. The electrical output of each generator is transferred underground to the bus bar by four, single-conductor cables. The bus bar itself consists of four cables stretched between two 4 by 4 posts. A secondary rack mounted on each post serves as an insulating support for the bus-bar cables. Two switches, one for each feeder line, are mounted above the bus bar. A wooden platform provides an insulating medium for operating personnel.

Whether or not your bus-bar installation looks exactly like the one in figure 8-10, the point to remember is that the equipment should be properly secured and supported, and, where necessary, properly insulated. The size of wire will depend on the load current. No. 6, single-conductor cable should prove sufficient for the feeder lines, while No. 1 or 250,000 CM cable can be used for the bus bar. The switches

Rgurc 8-10.—But bar installation. 245
that control the output to each feeder line can either be of the fused knife switch or the circuit-breaker type. Be sure that the current rating of the fuses or trip element will provide adequate protection against excessive overloads or short circuits on the feeder lines. Also make certain that the components of the circuit breaker or switch (i. e., the switch blades and breaker

contacts) are capable of carrying the rated current and voltage of the feeder lines. In addition, make every effort to protect the switchgear and bus bar from the weather. Building a weatherproof canopy over the rack will insure uninterrupted service and protection for personnel.

MOVING THE PORTABLE GENERATOR INDOORS

Although advanced base portable generating plants are designed to be operated outdoors, prolonged exposure to wind, rain, and snow wUl definitely shorten the life of the units. When conditions permit, the plants should be installed inside a building.

You won't find any predrawn plans covering the construction of a small advanced base generating station. It's an on-the-spot affair, the construction of which is determined by the equipment and material on hand plus your ingenuity, common sense, and ability to cooperate with other ratings. Before the Builder can construct the deck and generator foundations he'll have to get some information from you. This includes such things as the number of generators to be placed in the building; the dimensions of the generator's foundation; the location of the generators; the method of running the generator load cables between the generator and the bus bar and from the bus bar to the feeder system outside of the building; and the arrai^ement of the exhaust system.

Installation specifications covering some of the above-mentioned items can be obtained from the manufacturer's instruction book that accompanies each plant. The following hints and su^estions wiU help to fill the gaps:

1. Ventilation is an important factor to consider when installing the plants inside a building. Every internal-combustion engine is a heat engine. And, although it

is the heat that does the work, the exce^ amounts must be removed if the engine is to function properly. This can be accomplished by setting the engine's radiator grill near an opening in the wall and providing an outlet opening directly opposite the plant. In this manner, cool air can be sucked in and the hot air directed in a straight line outdoors. These openings can be shielded with. adjustable louvers to prevent the entrance of rain or snow. In addition, when operating in extremely cold weather, the temperature in the room can be controlled by simply closing a portion of the discharge opening. Additional doors or windows should be provided in the building if the plants are installed in localities where the summer temperatures exceed 80° F. at any time.

2. Working space is another factor that enters into the location of the generating plants. Be sure to provide suflBcierit space around each unit for repairs or disassembly and for easy access to the generator's control panel.

3. The carbon-monoxide gas present in the exhaust of the engine is extremely poisonous if allowed to collect in a closed room. Thus, means must be provided to discharge the engine's exhaust to the outdoors. This is done by extending the engine's exhaust pipe through the wall or roof of the building. If the length of the exhaust pipe is greater than 10 feet, be sure to use a pipe of not less than 1 ⅞ inches inside diameter. Support the exhaust pipe whenever possible and make certain that no obstruction or too many right-angle bends are allowed. Also, whenever possible, arrange the exhaust system so that the piping slopes away from the engine. In this way, condensation will be prevented from draining back into the cylinders. If the exhaust pipe should have to be installed in such a manner that loops or traps are necessary, a drain cock should be placed at the lowest point of the system. All joints must be perfectly tight and where the exhaust pipe passes through the wall care must be taken to prevent the

dischai^ed gas from returning along the outside of the pipe back into the building.

With these specifications in mind, the Builder will start the instaUation off by constructing the deck and generator foundations. Before he begins to pour concrete, however, make sure that all conduit is set in place wherever required and that a grounding network is prepared. After the concrete has set and the Builder gives the go-ahead signal, the generators can be installed on their foundations. In this job jou'U get your help from the Steelworker. He will use any means at his command (crane, A-frame, rollers) to set the units in place. Since the condition of the equipment is your responsibility, keep an eye open for any damage that might result while the units are being installed. The majority of portable plants are equipped with eyebolts and may be lifted easily from the top. However, you may find a unit that is only provided with four holes in the skid base, one near each corner. In lifting this type of plant, take care that none of the chains or lines press against any of the accessories attached to the engine.

Once the generating units have been set in place and the foundation bolts tightened, the Builder can proceed to erect the waUs and roof of the building. As soon as his job is complete you can finish the electrical installation. Grenerally, this consists of:

1. Setting up the bus bar, either on the wall or on a panel board as planned. The cables or copper rods serving as the bus bar should be well supported and insulated.

2. Installing the feeder disconnect switches if called for in the plans.

3. Installing the load cables in the conduit between each generator and the bus bar and making the connections at the generator terminals and at the bus bar. When connecting two or more sets in parallel, connect corresponding terminals together.

4. Installing the feeder wires between the bus bar and the outside lines via the feeder disconnect switches.

6. Making the ground connections between the neutral load cable and the grounding system. In most plants,

the neutral terminal T-O is connected by a link strap to generator frame. A short ground wire can then be connected from the frame to the grounding system. Caution Notes:

1. If the generating plant is to be moved from an extremely cold temperature into a warm room or building, the machine should first be covered completely with a piece of canvas or heavy paper. This cover should not be removed until the entire machine has had an opportunity to attain room temperature. Failure to do this wiU cause the machine to sweat, thereby reducing the insulation resistance of the windings.

2. Check the plant's instruction book before you remove the side housing panels. In some plants, the side panels are designed to help direct the cooling air where it is most needed and should be kept in place.

QUIZ

1. What are the basic components of an advanced base electric p>ower plant?

2. If you have acid with a specific gravity of 1.400, how many pai:ts of water and acid will you mix to obtain a solution with a specific gravity of 1.290?

3. W hat should you use to remove electrolyte spilled on the battery?

4. Before putting an alternator in service, what service and inspection steps should be followed?

6. What two factors must be considered in determining wire size on a

particular application? 6. What size exhaust pipe should you use if the exhaust is to be carried

more than 10 feet?

9«>«««nK<> 1 7

CHAPTER 9

CENTRAL ELECTRIC POWER STATIONS

SOME IMPORTANT CONSIDERATIONS

Large, permanent advanced bases require a huge amount of electric power. It is possible, of course, to supply this power from small, portable stations scattered tliroughout the base. However, it is much more advantageous in terms of reduced number of pereonnel and increased operating efficiency to supply the electric power from a permanent

CENTRAL ELECTRIC POWER STATION.

Basically, the equipment in the central power station is the same as that employed in the small advanced base stations. That is, there are one or more Diesel-tlriven generatoi'S to produce the electric power; switchboards to control the operation and outputs of the plants; and bus bars to collect and distribute the electric power. The size

and installation of the equipment is, however, another matter. Instead of portable skid-mounted plants rated at 440 volts, 75 kw., you'll find heavy duty stationarj'^ Diesel engines driving alternators rated at 4,160 volts, 300 kw, Instead of individually mounted control panels, you'll find switchboards controlling the operation of each stationary plant from one central location. And instead of an improvised, temporary layout, you'll find a preplanned permanent installation that is designed to produce electric power with a minimum of outages.

STATIONARY DIESEL-ELECTRIC POWER PLANTS

One type of stationary Diesel-electric power plant used in central power stations is shown in figure 9-1. At first glance it looks like a mighty complicated piece of machinery. Take another look, though, and you'll see that it's just a blown-up version of the familiar skid-mounted portable generating unit. It consists of a 2-cycle Diesel engine PRIME MOVER whosc 6 pistons transmit 450 horsepower to a crankshaft which rotates at a rated speed of 300 revolutions per minute. Directly coupled to the crankshaft is an ALTERNATOR that is capablo of converting the prime mover's 450 horsepower into 375 kilovolt-ampcres with a load power

PEOFSTAl BUtING

AlTERNAIOIt

rir WHEfl'

DIESEl ENGINE'

STAKTING CffNTROlS' ENGINE GOVERNOt

Fisore 9-1.—Fairbanks-Morse stationary Diesel-electric plant.

factor of 0.8. When the revolving field (rotor) of the alternator rotates at 300 r. p. m., it develops a 60-cycle, 3-phase, full-load voltage of 2,400/4,160 volts in the wye-connected stationary armature (stator). This electrical output is transferred out of the stator by four leads—a neutral wire and three main wires. A belted or direct-connected 10-kw., d-c generator supplies excitation current to the alternator's revolving field poles (rotor) via brushes and sUp rings. The main generator shaft, on which the rotor and exciter armature are carried, is supported by a pedestal bearing at the exciter end and by the engine's crankshaft at the flywheel end.

STATIONARY DIESEL-ELECTRIC PLANT PIPING SYSTEMS

The large stationary Diesel-electric plant is no different from the smaller portable plant with respect to the fundamental systems needed to ensure proper operation. Just as in the portable Diesel-electric plant, the stationary plant must have (1) a fuel-oil system, (2) a cooling-water system, (3) a lubricating-oil system, (4) an air-intake system, (5) an exhaust system, and (6) a starting system. The difference between the systems of the portable plant and stationary plant lies in the location of the parts of each system. Because it is designed to be moved easily from one spot to another, the portable plant will have all the parts of each system attached directly to the plant. For example, the radiator and cooling fan which form part of the cooling system are mounted right in front of the engine. Also, the oil pan which acts as a storage tank in the lubricating-oil system is bolted directly to the bottom of the crankcase. This is not the case, however, with the stationary Diesel-electric plant. Because of its size and operating characteristics, the stationary plant is separated from the parts which make up each of its systems. The location of these parts and their place in a particular piping system are discussed in the foUowing paragraphs.

The puel-oil system carries the fuel to the engine and provides a return path for the surplus fuel. Each stationary

engine will have its own fuel-oil supply tank. For ease in filling and for fire prevention purposes, the fuel tank is usually located outside the building. It is placed in a horizontal position in a covered pit as shown in figure 9-2. A suction pipe and an overflow pipe connect the tank to the engine. Inside the building, the pipes are brought over to the engine in concrete trenches. In some instances a lai^e main storage tank is installed in addition to each engine's fuel supply tank. This main tank is placed at a level from

Fisurc 9-2.—Fuel piping diasram.

which the fuel will flow by gravity into the smaller supply tanks. With this setup, the engine's fuel supply tank is designated as a day tank since its capacity is such as to ensure the operation of the Diesel engine for at least 8 hours.

The COOLING-WATER SYSTEM removes the excess heat from the engine by circulating soft cooling water through the engine water jackets. There are two systems used to recool the cooling water before it is sent back thiough the engines— the DIRECT SYSTEM and the indirect system. In the

direct system, the heated cooling water is piped to a spray or cooling tower where its temperature is lowered both by exposure to open air and by evaporation. The recooled water is then pumped back through the engine water jackets to complete the cycle. In the indirect system, the cooling water is recooled by means of a heat exchanger or a radiator. Raw water is used as the cooling medium in the heat exchanger while a fan must be provided to force air through the radiator. The indirect system is preferred to the direct system because there is less loss of cooling water by ev apora-tion and consequently less makeup water required. In both systems, the cooling units and their accessories are usually located outside of the building. The piping system that carries the cooling water from the cooling unit to the engine and back to the cooling unit may be placed in concrete trenches or on overhead supports. Figure 9-3 shows two radiator coolers located outside of a central power station that houses two Diesel-electric power plants. Mounted on the back of each unit is a large fan. The small bo.xes

Fisure 9-3.—Radiator coolins units. 255

located on the upper right-hand corner of each radiator are the expansion and makeup tanks.

The LUBRICATING OIL SYSTEM has the job of circulating clean oil through the moving parts of the engine. The clean-oil storage tank and pressure type oil filter are mounted near the engine as shown in figure 9-4. All other parts of the system are either attached to or incorporated in the the engine itself; so installation requires only the connecting of pipe between the used-oil sump and the filter, and another pipe between the clean-oil sump and the storage tank.

The AIR-INTAKE SYSTEM suppHcs clcan air for combustion purposes and for cleaning out (scavenging) the exhaust gases from the cylinders. The air may be drawn into the engine by one of two routes: (a) from inside the building through a screen located on top of the engine base, or (6) from outside the building through an underground conduit into the engine base. Route (6) is the preferred method and when it is used route (a) is sealed off. Figure 9-5 shows the method of bringing in the scavenging air from outside the building. The conduit is made at the time the enghie

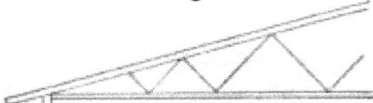

Figure 9-5.—^Typical air-intake tyitem.

foundation is built. Notice that the outer end of the conduit pipe has been extended above the ground level to keep water out, and a suitable screen covering is provided to keep dirt from being drawn in.

The EXHAUST SYSTEM providcs a means of removing the scavenged exhaust gases from the engine cylinders to the atmosphere. There are two possible arrangements of the exhaust system. In one arrangement (fig. 9-6) a concrete conduit extends lengthwise and adjacent to the engine foundation. Separate pipes from each engine cylinder lead into this conduit which in turn is vented to the atmosphere through a stack. If this method is not practicable another arrangement can be used in which exhaust stacks connected to individual exhaust pots are run through i the side waU or roof of the building.

The STARTING SYSTEM providcs and controls the energy required to start the engine. In large, stationary Diesel engines, this enei^y is in the form of compressed air. The starting sytem, itself, consists of an air start mechanism

Figure 9-6.—Undersround exhaust system. 258

on the engine for distributing and admitting the compressed air to the proper cyHnders, an auxiliary air compressor pump to build up the air pressure, steel tanks to store the compressed air, and the necessary, piping to transfer the compressed air from the tanks to the engine. The air compressor pump and steel tanks are usually placed in one corner of the power station. The piping system is run in concrete trenches to the engines. Figure 9-7 shows an actual air compressor and tanks installation.

Figure 9-7.—Air compressor and tanks installation.

CENTRAL POWER STATION SWITCHGEAR

The terms, "switcliboard," "switching equipment," and "switchgear" are often times confused with one another. If you need straightening out on these three terms then check the following definitions:

SWITCHBOARDS—are usually defined as the panels or control desks on which the meters, relays, instrument switches, and control handles are mounted.

SWITCHING EQUIPMENT—^is usually Understood to mean the cux'uit breakers, switches, and similar equipment in the

main circuit including current and potential transformers, bus bars, connectors, etc.

swiTCHGEAR—is fi coordinated assembly of switchboards and switching e(iuipment necessary for the control and distribution of the electric power produced in the central power station.

You won't liaye much trouble spotting the switchgcar in a central power station. It's a big brother to the type that's found on each portable advanced base generating plant. You'll notice, however, that its location has been changed. Instead of being made an integral part of the plant itself the switchgcar is moved to a central location and combined with the switchgcar that controls each of the other plants in the power station. The tie-in between each plant and its respective switchgcar is made by cables laid in conduit. The advantage of this centralized setup is obvious—the output of each plant can be controlled, collected, and distributed from one location by one operator.

Fisure 9-8.—Switchsear for a central power station.

In fi^jure 9-8 yoirre looking at the switchgoar in a central power station containing two Diesel-electric plants. Notice that the vertical switchboaids of the right-hand and center switchgear are similar in every respect. Each controls the output of one of the Diesel-electric plants. The third switch-gear unit on the left is an au.xiliary unit that carries meters and switches common to both plants.

Figure 9-9 is a closeup of the switchboard of one of the two similar switchgear units. Mounted on the panel is the following equipment:

St cnii M »ui
SO
(onaK IMP WW lORiA

laaint \ni!i miwni iff
U(JIf» Itii IMP!"'

Fisure 9-9.—Swikhboard view of a switchgear unit.

D. c. AMMETER.—This meter indicates the curreDt output of the exciter which is the input to the alternator field.

Frequency indicator. —This meter indicates the frequenci, in cycles per second, of the generated power.

A. c. wattmeter. —This meter indicates the power output in kilowatts of all three phases of the power plant.

A. c. ammeter. —This meter indicates the current output of the alternator.

Ammeter switch. —This three-position switch connects the a-c ammeter into each phase line of the alternator so that the current in each phase line can be read by one met^r.

A. c. voltmeter. —This meter indicates the phase to phase voltage of the alternator's output.

Voltmeter switch. —This four-position switch connects the a-c voltmeter between each phase line of the alternator.

Voltage regulator unit, —This control unit maintains the voltage of the power plant at a predeteimined value by automatically raising or lowering the excitation voltage supplied to the alternator as necessary to maintain the preset voltage.

Voltage regulator switch. —^The position of this swit<?h determines^ whether the voltage output of the alternator is to be regulated automatically or manually.

Synchronizing lamps. —These lamps, when placed in the circuit, permit a visual comparison of the frequency and phase relationship of the incoming generator power with that of the power already present on the bus.

Synchronizing switch. —This switch is used to place the synchronizing lamps in the circuit when the alternator is to be paralleled with another unit.

Exciter field rheostat. —This rheostat located in the exciter's field circuit permits a manual control of the excitation of the exciter field and thereby a manual control of the a-c voltage of the alternator.

Field switch. —This switch is in the cii'cuit between the exciter and the alternator field. With this switch open, the only voltage that can be generated by the main generator is the voltage due to the residual magnetism of the generator field structure. Main circuit breaker. —This breaker switch (visible in lower portion of switchboard, figure 9-8) is connected in the main power circuit between the generator and the outgoing bus. Its purpose is to connect or isolate the power plant generator to or from the outgoing bus. Elapsed time meter. —This meter records the operating time in hours when the unit is generating. Governor motor control switch. —This switch permits a remote control of the governor speed setting of the Diesel prime mover. Anti-motoring relay. —This relay is designed to open the main circuit breaker when the power flow is reversed; i. e., when the alternator is being driven as a motor. This may occur in parallel operation with another plant if the governor of the Diesel engine is tending to hold too low an engine speed. . Generator disconnect switch. —This is an interlock safety switch which must be operated before the main circuit breaker can be closed or open. Cooling fan motor control. —This push button switch starts or stops the cooling fan mounted on the radiator units outside the building. (See fig. 9-3.) The meters, control knobs, and switch handles that arc mounted on the front panel of the switchgear imit either control or are controlled by other equipment located behind the switchboard. The output cables from the plant's alternator enter the back of the switchgear unit and connect to the input terminals of the main circuit breaker. The output terminals of the circuit breaker are connected to the main bus bars which serve as the collecting and distributing point for all power entering or leaving the power station. Connected to or mounted on the bus bars are the current and potential transformers that supply stepped-down voltages and currents to the meters, relays, and to the voltage regulator unit. In addition to this switching equipment, protective equipment such as lightning arrestors and capacitors are also located in back of the switchboard.

They are connected from each hne of the main bus to the ground bus and protect the power plant in the event that hghtniiig should strike one or more of the feeder lines connected to the plant.

All of the equipment located behind the switchboard is rigidly mounted on insulated supports. The wiring for the low voltage and low current circuits is usually made up of fire-resistant covered wire arranged in a neat and systematic manner. Interconnection of bus bars between adjacent switchgear is made by cable across the top of the switchgcar. These connections are clearly visible in figure 9-8.

METAL-CLAD SWITCHGEAR

Figure 9-10 shows an actual installation of metal-clad switchgear in an advanced base central power station. It is the newest type of switchgear and the one you will probably

Fisuie 9-10.—Metal-clad switchgear installation. 264

operate most frequently. Basically, metal-clad switch-gear is no different from the older type of switchgear shown in figure 9-8. That is, its fundamental purpose is still to collect, control, and distribute the electrical enei^ developed in the central power station. And you'll find that it needs the same switching equipment, meters, and relays to do the job. It differs only in its construction and internal arrangement of equipment.

KTMCUO

nnrcMGUi

MKHmsanKUi iwii no. i uwi wo. i umi mo. 3 umi no. <

suui noncnoi hneutoi Excini contioi Fcani m

(ONTIOl

Figure 9-11.—Comparison of metal-clad switchscar to older-type switchgear.

Figure 9-11 will give you a clearer picture of the differences in construction between the older type switchgear and the metal-clad switchgear. Each switchgear represents the minimum equipment required for the control of one electric generating plant. Notice that the older type switchgear on the left is a one-unit affair. That is, aU of the switchboard equipment, is on one panel and the switching equipment is contained in one compartment. Now take a look at the metal-clad switchgear. Instead of one unit, there are four separate units attached together. Each unit contains meters and/or equipment necessary for the control of a definite phase of the generator's output. Unit number 1 is

for sui^e protection. It contains a set of main bus bars, capacitors and lightning arrestors, and a potential transformer for the bus potential voltmeter. Unit number 2 is the main control unit for the a-c generator. It contains an oQ-immersed main circuit breaker, a set of main bus bars, and the current and potential transformers necessary for the operation of the meters, relay instruments, and indicating lamps on its front panel. Unit number 3 is the exciter control unit. It contains a set of main bus bars, a voltage regulator unit, exciter field rheostat, and potential transformers for the operation of relays and meters. It also is the unit from which the auxiliary station power is taken via 3 current-limiting fuses. Unit number 4 is the feeder circuit unit from which the station power collected on the bus bars is sent out of the station through feeder wires. This unit contains a set of main bus bars, a feeder circuit breaker, current and potential transformers, plus the necessary meters, relays, and switches. Finally, there is the swinging instrument panel attached to the top of unit number 1. Its construction is such that all meters on its panelface can be easily seen by the operator who is handling the controls on the switchgear. These meters include an a-c voltmeter that indicates bus potential, and a synchroscope with two synchronizing lamps.

Each unit is factory assembled, wired, and tested before shipment. Thus, coordinating each unit into one assembled switchgear amounts to mounting the units on their foundation and connecting main bus bars and control circuits. There will be one assembled switchgear for each generating plant in the station. If you'll check back to figure 9-10 you'll see that there are five complete switchgear assemblies. This indicates that there are five Diesel-electric plants in this particular power station. Notice that only one surge protection unit and one swinging panel is required for all five switchgear assemblies. Also notice the method by which the feeder lines leave the switchgear and the building on their way out to the distribution system. Only three feeder lines are shown; two more are to be installed.

METAL-CLAD SWITCHGEAR CONSTRUCTION

The internal construction of metal-clad switchgear is such as to insure complete protection of operating personnel. Circuit breakers and other associated switching equipment such as transformers, insulated busses, and connections are placed in separate, grounded, metal compartments. Figure 9-12 is a cut-away view of a metal-clad switchgear unit.

Fisure 9-12.—Internal construction of mctal-clad switchgear.

The metal-enclosed compartments are clearly visible. Notice that the front panel of the switchgear can be swung open, providing access to the meters and relay connections and to the circuit breaker without exposing the operator to live circuits in adjacent compartments. The neat and systematic arrangement of connecting wires can also be seen.

Access to the current transformers and main bus is provided at the rear of the switchgear. If you want to inspect the current transformers and input cable connections, you simply Temove the rear barrier by unscrewing a few bolts (fig. 9-13).

Figure 9-13.—Removing rear barrier For inspection of current traniformcrs

If further inspection of the main bus compartment is necessary, additional steel panels are removed as shown in figure 9-14.

The method by which the bus bars extend from one switch-gear compartment to another compartment can be seen in figure 9-15. A close-fitting insulating plate not only supports the bus bars as they extend through the adjacent compartment, but also prevents gas migration. The busses are made of copper strap insulated with preformed insulating tubing on straight portions and varnished cambric insulating tape at bends. Bus taps are made with high-pressure connection bai s and all tap surfaces are silvered to ensure a low-rcsistant joint. The method of hisulating bus taps is shown in figure 9-15. A preformed insulating box is fitted around the tap and clamped into place. Tlie two holes in the box provide a means of filling it with an insulating compound.

Figure 9-14.—Removing barriers of main bus compartment.

METAL-CLAD SWITCHGEAR CIRCUIT BREAKERS

If there's any doubt in your mind as to whether the switch-gear equipment you are working on is of the metal-clad type, just take a look at the circuit-breaker equipment. If you find that the circuit breaker is both removable and interchangeable as a unit, then it is metal-clad switchgear.

Figure 9-16 shows the horizoxtal drawout type circuit breaker employed in Westinghouse metal-clad switchgear equipment. This circuit breaker is an air circuit breaker. That is, the breaker contacts open and close in air. Arcs that form during the opening of the breaker contacts are directed upward into a multi-grid arc chute where they are extinguished by the action of a magnetic field. This is known as a DE-ION breaker. Each breaker contact is covered by an arc chute. In figure 9-16, two of the arc chutes have been removed.

Fi3ure 9-15.—Insulafing the bus tap.

Notice that each breaker contact has a horizontal extension stud that extends through the front steel panel of the breaker unit (fig. 9-16.) These primary disconnects provide the means by which the incoming power to the switchgear is transferred through the circuit breaker and over to the bus bars. How this is accomplished is showii in figme 9-17. (This illustration is a side view of the switch gear compartments—if you need a little orientation check back to figure 9-12.) Notice (fig. 9-17) that the three main wires from an a-c generator enter the rear compartment of the switchgear, travel through the current transformers, and connect to three separate horizontal sockets. Immediately below these sockets are three more horizontal sockets, each connected to one of the three bus bars. When the circuit breaker unit is pushed into place, the six primary disconnects enter and make firm contact with their respective sockets. The circuit is * then completed between incoming lines and bus bars when the breaker contacts are closed.

The steps in installing a horizontal-drawout circuit breaker are illustrated in figure 9-18. The first step consists of lining up the circuit breaker unit's wheels on the track in the switch-gear compartment. In the second step, the circuit breaker is pushed into the compartment until stopped by the drive-in device at the test position. In the third step, the circuit breaker is moved from the test position to the operating position. This is accomplished by a worm gear device that is operated by a removable hand crank. A metal shutter, that normally covers the incoming line and bus bar sockets, opens automatically as the breaker advances to the operating position. At the same time that the breaker primary disconnects enter and make contact with these sockets, a secondary connecting device will automatically engage. This connecting device, consisting of multiple plug and socket contacts, provides connections for the contol circuits between the circuit breaker unit and the switchgear housing. Removing the circuit breaker, of course, is just the reverse process of putting it in.

The safety features employed in the removable circuit breaker will prevent accidents that normally result from for-getfulness on the part of the operator. For exaihple, if you try to crank a closed breaker into or out of its operating position, an interlock will prevent crank motion. Also, the breaker cannot be closed while it is traveling from test to operating position. Another safety feature is the provision for grounding the circuit breaker frame. A ground bus extends throughout the length of the metal-clad assembly.

I

Figure 9-18.—Installing a horiiontal-drawout type circuit breaker.

Each switchgear housing is grounded directly to this bus. Wlien the circuit breaker is placed in the switchgear compartment, a grounding jaw attached to the breaker frame makes contact with a knife blade attached to the ground bus. This grounding connection is automatically made before the circuit breaker is in the operating position.

The VERTICAL-LIFT circuit breaker is another type of removable circuit breaker employed in metal-clad switchgear. As its name implies, it is lifted vertically into engagement rather than through a horizontal motion as is the case with the horizontal-drawout circuit breaker. And, whereas the

273

horizoiital-diawout breaker is mounted permanently on its own triu'k, the vertieal-lift breaker unit is placed on a "dolly*' transfer truck that is removed after the breaker is lifted into position. Figure 9-19 shows the three steps involved in the installation of a vertical-lift circuit breaker. In step 1, the circuit breaker is rolled into the switchgear housing. Floor

Figure 9-19.—Installins a vertical-lift type circuit breaker.

I

guides and position-limit stops provide proper alinement of the transfer truck and breaker. In step 2, the breaker is , raised manually by a screw-jack mechanism operated by a removable crank (some equipment has motor-driven elevating mechanism). In step 3, the transfer truck has been withdrawn and the breaker elevated into the final operating position. Mechanical interlocks are provided, of course, to ensure safe sequence of operation. The vertical-lift type of circuit breaker is employed in General Electric metal-clad switchgear equipment.

CENTRAL POWER STATION CONSTRUCTION

Although the ultimate output of the central power station is electric power, don't get the idea that you'U be the only one involved in its construction. Actually you'll be just one part of a skilled team of many ratings, all working under the direct supervision of CEC officers. And it's not a hit-or-miss affair, either. Such factors as the number of plants to be installed, the type of building to be erected, the location of the station, expected weather conditions, available equipment, and many other important considerations are taken into account. Blueprints covering the mechanical and architectural construction details and electrical drawings showing the location of electrical equipment and their connections are prepared. Helpful guides are also furnished in Advanced Base Drawings, NavDocks P-140, plus the manufacturer's instruction book that accompanies each piece of equipment.

Generally, you'll find the following phases involved in the construction of a central power station.

1. The assembly and erection of the Diesel-electric plants.

2. The installation of underground mechanical piping systems and electrical conduit systems.

3. The installation of mechanical equipment and electrical equipment.

4. The grounding of the station's equipment.

ASSEMBLY AND ERECTION OF DIESEL-ELEaRIC PLANTS

Each Diesel-electric plant will be shipped partly disassembled. That is, each major part such as the Diesel-engine stator, rotor, fly wheel, and extension shaft mil be crated separately. These crates are the parts' best protection, and therefore the equipment should not be unpacked until just before it is installed.

Due to their size, Diesel-electric plants are usually erected before the walls of the station are built. Therefore, the first thing to be constructed, even before the deck is poured will be the plants' foundations. Made of concrete, each foundation will conform to the dimensions of the plant's base and will include a pit at the generator end to accommodate the extra-lai^e diameter flywheel and rotor. Prior to the pouring of concrete all conduit is placed in position. This conduit will provide a protected path for the main generator and exciter wires through the foundation wall.

You can get a lot of information from f^ure 9—20. First of all, you can see the general appearance of the foundation. Second, you can see a t\T)ical method of moving the Diesel prime mover onto the foundation. Notice that the prime mover is supported on wooden cribbing placed aloi^ide the foundation. Steel cables attached between the prime mover and a power winch (located out of view), supply thq pulling force necessary to slide the prime mover over the way-planks onto the foundation. Actually, the prime mover is not placed directly in contact with the foundation, but is supported by levehng screws and steel bearing plates. A good machinist's level is then placed at various positions around the engine's base, and the leveling screws adjusted until an accurate crosswise and lengthwise level is obtained.

After the engine base has been leveled, the fl3rwheel, extension shaft, alternator, exciter, and pedestal bearing are installed and alined. These are all heavy parts that call for some mechanical advantage in the form of a chain hoist. Support for the hoist is obtained as shown in figure 9-21. This is a temporary measure and is replaced later by a permanent hoist that runs the length of the entire plant. Notice that the flywheel has already been lifted and secured into position and the bottom half of the bearing bushing is in place. The rotor mounted on the extension shaft is fitted next. It is lowered into place with the shaft resting in the lower half of the bearing at one end and bolted to the flywheel hub at the other end. Then the stator is placed in position around the rotor and bolted to its foundation. This step is shown in figure 9-22. Every precaution should be taken

Figure 9-21.—Temporary chain hoiit arrangemenh

during this opcMation to prevent injury to both the stator and rotor wintlings. A piece of one-sixteenth-inch sheet steel phiced in the lower half of the stator bore will protect the core and offer a smooth surface for the rotor to slide on. The upper half of the pedestal bearing is then assembled. To complete the job it is necessary to recheck the alinement of the engine and extension shaft and pour cement under the engine base, the pedestal bearing sole plate, and the stator foundation. After the cement has set, the leveling screws and beating plates are removed and the main foundation bolts tightened.

Figure 9-22.—Lifting the stator into position.

INSTALLATION OF PIPING AND CONDUIT

\Mienever possible, the fuel, water, and air pipes and the electi'ical conduit are laid below the level of the floor. This not only improves the general appearance of the station but is also a definite safetv feature. This work is done, of couree, before the concrete deck of the station is poured.

The possibility of a break in the piping systems (fuel, air, water) is always present. Therefore, the pipes are made accessible by laying them in channels covered by steel or wooden floor plates. These channels arc dug about 5-inches deep and 12- to 18-inches wide. Theij-location and exact dimensions are obtained from the floor plan blueprint.

The electrical conduit ofl'ers protection for the cables that run between each Diesel-electric plant and the switchgear and other equipment in the station. They are laid directly in the ground and run with a minimum of bends. Where a

90** change in direction is required on long runs, handholes are provided to facilitate the pulling in of the cable. The exact location of the conduit runs, the conduit coupling details, the size of conduit to be installed, and other pertinent information is completely furnished on the electrical plan.

INSTALLATION OF MECHANICAL AND ELEaRICAL
EQUIPMENT

Another construction phase involves the installation of the mechanical and electrical equipment necessary for efficient station operation. The mechanical equipment includes such units as the air compressor and the air, fuel, and oil tanks. The location, construetlor details, and piping connections of each piece of mechanical equipment is furnished by the station's floor plan and the manufactiuer's drawings.

The electrical equipment falls into two classes; those that function in the high-voltage output circuit of the station, and those that make up the low-voltage secondary circuits in the station. The high-voltage equipment is primarily composed of the swiTCHGEAR and its accessories. The secondary circuit equipment includes the transformer bank that reduces the station's high voltage to the low voltage required for the secondary circuits, the power panelboard that distributes the low voltage to the secondary circuits in the station, the electric motors that drive the mechanical equipment (evaporator fans, evaporator pimips, air compressor pumps, and jacket water pumps), the lights that provide illiunina-tion for the station, the receptacles that provide convenient power outlets, and the switches that control both the lighting and power circuits.

Your guides in the installation of the station's electrical equipment will be the lighting and power plans and other detailed information furnished by the manufacturers of the equipment. Typical of such lighting and power plans is the one shown in figure 9-23. It represents half of the electrical plan of a central power station that contains three Diesel-electric plants. The building is a 40- by 100-foot quonset

type. Notice the information available on the print. For example, you can see the type of switchgear to be installed the number of units used, their physical location, and a note pertaining to the method of switchgear installation. Other information to fill in the gaps will be found in detailed drawings of other views of the switchgear, wiring diagrams, and the manufacturers* instruction books. Details on the power panelboard are also available from the plan. This includes the physical location of the panelboard; a reference note for moimting details; and the location, type, and number of cables that make up the secondary lighting and power circuits. Again, other information such as the type of panelboard used, its wiring circuit, and the methods of supporting the cable on the roof and waD«^of the building will be found on other detailed drawings. Information on the electric motors and their starting switches is also included on the electrical plan (fig. 9-23). You can see, for example, that two air-compressor motors are located in the upper right-hand corner of the station, a jacket water pump is placed near the wall opposite the plant, and an evaporator fan motor is set in a shed outside the building wall. The starter switches for each motor are mounted on the wall and connected electrically to the motors by three, single-conductor, super-sheath. No. 6 cables placed in 2-inch conduit. Finally, the droplights and receptacles are shown in their physical location on the electrical plan. The wiring connections for their circuits can be obtained from lighting and receptacle circuit drawings that are standardized by the Bureau of Yards and Docks for all 40- by 100-foot quonset buildings.

GROUNDING THE STATION'S EQUIPMENT

For the safety of personnel and the protection of equipment, all metal parts of the station equipment which normally do not carry electric current, must be effectively groimded. This equipment includes the following:

1. The frames or metal base of each electric motor and generator.

NON-nntD
SWITCH

TI4T
\J,IO»M t<CiTt« ^
JACKET WATtS (VM»
m 1
oooo
OdSEL OlNtRATO* N«l
I5f—^
OAVtLtCT«tC *i« . IJ
o /
CAtitS too vOIT

LCAVt AN ASEA LASOEa tV *' ON AIL ».r>€S 0» V«1TCM MA* WITH A 2-aiCCSS IN flCX>« FOB SETTING rjCHANNEL laCH^ LCvlt OaOuT^N AROUND CHANNtl. laOMS TO FIN'SMIO

rtooft HtiOH^

UfTAL CLAD

SiMTCH c»c*a

n-t'

4-(i)

TtANtKWMEBS

5-10 KVA

NDOMEP PANEL.K)« MOUNTINO EXTAuS.StE 0x0 N< >05*«4

"B"«-l

6-lt*«-*ao VOLT »uafa-SHCATM -c.^» O

2-il:*«-«oo voiT

»UPt*-W<UTH

«-»OOVOLT SUPi«-5HIATH CAALI

Fisure 9-23.—Typical station lishtins and power plant.

- tVAPOaATOB

m

lue. OIL

/TANK

,EVAI>0II*10<

/TANK

OROUNO counter *

eclioiNO sTttt

S75 KVA DlESeL GENERATOR Nil

—rr — H-

Aia COM-

pausoas LJ IJl

i.

LL SaANCMES TO 51

*t uat bOPxa UCOT

AS OTHERWISE NOTES

METAL CLAO / SWITCH C)E*a

4

S-IO KVA ^_TRANJ»0«ME«*

-J Jj!iLl>i&U a^g---iLJ

POWER PANEL

LETTERS A,6 & C INDICATE OaouNDINO

DETAILS.

THIS ORAWINO

PLAN OF 6ROUN0IM6 SYSTEM

SCALE '/&'• l-O'

Figure 9-24.—Grounding plan. 282
2. The metal cabinet of the power panelboard.
3. The neutral bus in the metal-dad switchgear.
4. The steel tanks of the transformers.
5. The steel posts that support the chain hoists over each plant.
6. The building frame, if it is of metal.

Since this equipment is scattered throughout the station, a grounding system must be installed. Essentially, this consists of a main belt of .bare stranded copper cable that encircles the station. Branch cables are connected between the equipment and the main cable which is grounded to earth at frequent intervals throughout its run.

The grounding system is placed under the floor of the station. Therefore, the laying of the cable, the driving of the ground rods, and provisions to ground the equipment after it is installed must all be done before the concrete deck is poured. All details concerning the location and connections of the parts of the grounding system will be found in the GROUNDING PLAN for the particular station you are helping to construct. Typical of these grounding plans is the one shown in figure 9-24. The dotted hues represent the grounding cables. Notice that No. 1/0 bare stranded copper cable is used for the main belt while No. 6 bare stranded copper cable is designated for the branches— except as otherwise noted. With the plan as a guide, the cables are laid in a slack fashion along the prescribed routes. Branch cables are placed at all points marked "A" on the groucding plan. The grounding plan also specifies the spots where you are to drive the grounding rods. These points are marked with the letter "R."

With grounding cables and ground rods in place, the next step consists of connecting them together. The ground rods are connected to the main belt cable with a solderless clamp as shown in detail view "R," figm-e 9-25. Similarly, the branch cables are tapped to the main cable as shown in detaU view "A."

The points on the grounding plan marked with the letter "C" designate the connection of the branch groimd cables to

their respective equipment. Since the branch cables must come up through the concrete deck in order to connect to the equipment, some provision must be made to keep the cables in position while the deck is being poured. How this is accomplished is shown in detail "C," figure 9-25. A length of one-inch conduit' is supported in a vertical position near

|—rxio'-o*
, J GROUND ROD
FRONT VIEW SIDE VIEW
DETAIL
HO tCAU

•6 BARt STR"(COPPER BRANCH
FRONT VIEW
DETAIL
NO SCALC
SIDE VIEW
@ -20 MACHtNE BOLT ® M LOCK WASHER
6ASE OF MACHINE OR EQUIPMENT

TAP FOR Xi-TOX I* LONG MACHINE BOLT AND CLEAN SURFACfS
I" CONDUIT
FILL WITH PITCH TOP OF SIAe
**A> 6A«E WIRE MAIN ORO UIN0^~Ny^ ^
SEE DETAIL 'A
DETAIL "C*
NO SCALt
^"■^ 6RAN04-FOR Size tft ■ PLAN ON THIS SHEET

Figure 9-25.—Groundinj connection details.

the contemplated site of the equipment. The branch cable is passed up through this conduit and temporarily lashed in place during the pouring of the concrete deck. After the concrete has set and the equipment has been installed, the branch cable is connected to the metal base of the equip* ment by means of a lug attachment (detail "C," fig. 9-25).

POWER PLANT OPERATION

Constructing the central power station is just one phase of your job. When that's finished you'll be expected to supervise the activities of the operating personnel of the station. In this respect, your supervision should be directed toward one ultimate goal—to maintain a continuous and adequate flow of electric power out of the station. This can be accomplished only if you have:

1. A thorough knowledge of how to operate and maintain the equipment in the station;
2. A complete understanding of the station's electrical system as a whole.

Obviously, a thorough knowledge of how to operate and maintain the specific equipment found in the station to which you are assigned cannot be included in this chapter. Only the general procedures can be given. It will be up to you to supplement these general procedures with the specific instructions given in the manufacturers' instruction books furnished with each piece of equipment. Similarly, familiarity with the station's electrical system as a whole can only be gained by a study of information relating specifically to that installation. This information can be found to some e.xtent in the manufacturers' instruction books but the greater part of it will be obtained from the station's electrical plans and wiring diagrams. Remember, however, that a study of the electrical plans and diagrams must be supplemented with an actual study of the station's system itself. In that way the generators, switchgear, cables, and other electrical equipment are not merely symbols on a plan but physical objects whose location is definitely known, and whose functions and relations to the rest of the system are thoroughly understood.

THE GENERATOR WATCH

A series of around-the-clock watches must be scheduled if the station is to supply a continuous and adequate amount of power. Depending on the number of operating personnel available, the watches are evenly divided over the 24-hour period. Although 6-hour watches are common practice they may be stretched to 8-hour stints without working undue hardships on the part of the operators. Watches exceeding 8 hours should be avoided unless emergency conditions dictate their use.

The duties involved in a generator watch can be grouped into three main categories: (1) operating the station's equipment, (2) maintaining the station's equipment and (3) keeping a daily operating log. Operating and maintaining the station's equipment will be covered in succeeding sections in this chapter, so for the present you can concentrate on the importance of the third duty of the station operator—keeping a daily operating log.

The station log serves three purposes. First, it will keep the man in charge of the watch on his toes. He is not likely to relax his vigilance, especially diu-ing late watch shifts, if he must make an hourly recording of all gages and instruments. Second, since the number of operating hours is recorded in the station's log, the log serves as a basis for determining when a certain piece of electrical equipment is ready for inspection and maintenance. And third, the station log can be used in conjunction with previous logs to spot gradual changes in equipment condition which ordinarily are difficult to detect in the day-to-day operation of the station. A suggested station log is shown in figure 9-26.

The physical appearance of the station is a reflection of your ability to supervise the activities of the watch standers. A clean and orderly power station is usually well run and well maintained. Keep that thought in mind when you furnish instructions to the operators concerning the general care of the station. Impress on them the importance of keeping tools in their proper place when not in use. The same precaution applies to any auxiliary equipment which is not normally used in the regular operation of the station. Clean waste and oily waste should be kept in separate containers. Emptying the oily waste container at least once a day will reduce fire hazards. The amount of care given the station floor will be governed by its character. Generally.

Figure 9-26.—Station log.

it should bo swept down each watch. Any oil or grease that is tracked around the floor should be removed at once.

SINGLE PLANT OPERATION

Connecting an electric plant to a deenergized l)us involves two general phases—(1) starting the Diesel engine and bringing it up to rated speed under control of the governor and (2) operating the switchboard controls to bring the generator's power onto the bus. Remember again that the procedures given here are of necessity general in nature. Detailed instructions for the operation of the particular plant you are working on will be found in the manufacturer's instruction book accompanjnng the equipment. Where a variation is noted between the specific instructions given in the manufacturer's book and the general instructions given in this section, the manufacturer's instructions will, of course, take precedence.

The engine controls and operating gages are usually placed on the engine itself. A general view of the controls and gage instruments of a General Motors Diesel engine is shown in figure 9-27. The function of the gages mounted on the

panel board are indicated by their identifying titles and are self-explanatory. The engine controls consist of the starting AIR LEVER which controls the valve that admits starting air to the engine's cyUnders, the engine operating lever which controls the engine speed while starting, stopping, or when under emergency control, and the speed control KNOB which controls the engine speed through the governor.

Figure 9-28 gives you a close-up view of the engine operating lever and its component parts. One such part is the pawl-actuating lever. When the handle of the pawl actuating lever is moved so that it points toward the operator, the governor becomes inoperative, and the operator can

Fisure 9-27.—Diesel ensine controls and gases.

manually control the engine speed by moving the operating LEVER HAiNDLE. This is the normal condition when starting the engine. A vernier knob provides accurate speed adjustment when operating under emergency or manual control. The LATCH LEVER is connected to a latch that holds the operating lever handle at the exact spot to which it is set. Wlien you grasp the operating lever handle, the latch lever is near enough to be conveniently controlled by your fingers. With the pawl actuating lever turned away from the operator and the operating lever-handle latched in the "gov. run" position, the governor has complete control of the engine

Figure 9-28.—Engine operating lever details.

speed. This is the normal condition after the engine is started.

Fif^ure 9-29 shows the Diesel engine in the initial process of heing started. Notiee the positions of each hand—one grasping the operating lever handle, the other holding the starting air valve lever. (Caution note: before starting the engine, make certain that hoth the circuit breaker and field switch on the plant's switchboard arc in the off position). Generally, starting the engine involves the following steps:

Fisuie 9-29.—Startlns the Diesel ensine.

1. Set the pawl-actuating lever for manual control and latch the operating lever in the inj. full position.

2. Admit the starting air to the engine's cylinders by pulling down on the starting air lever. The engine should fire within 10 seconds. Release the starting air lever when the engine fires.

3. Check the lubricating oil pressure. If pressure is not indicated on the gage within 10 seconds stop the engine.

4. Release the pawl-actuating lever and latch the operating lever in the gov. rux position. The governor now has full control of the engine.

5. Check the engine speed on the r. p. m. indicator gage.

This particular plant must run at a rated speed of 600 r. p. m. to produce a 60-cycle voltage output. If the engine speed is not 600 r. p. m., turn the speed control knob until the dial needle on the r. p. m. gage points to 600 (fig. 9-30).

Figure 9-30.—Using the governor control knob to obtain rated engine speed.

Check the lubricating oil gage for proper oil pressure. Check the fuel oil gage for proper fuel pressure. Check the temperature gage for proper circulating water temperature. Make other checks as outlined in manufacturer's instruction book for the particular plant you are operating.

Fisure 9-31.—Supplyins excitation voltage to alternator's field.

With the engine under governor control and operating at rated speed you are ready for the next broad phase of plant operation—operating the switchboard controls. The general steps involved are given below:

7. ^lake certain that the voltage regulator switch is in the OFF or manual control position and that the voltmeter switch is set to permit the a-c voltmeter to read line voltage in one phase. Turn the exciter field rheostat handle as far clockwise as it will go. This inserts maximum resistance in the shunt field of the exciter.

Figure 9-32.—Adjusting alternator output voltage for rated value.

8. Watching the a-c voltmeter, throw the fieldswntch to the ON position (fig. 9-31). This action makes the ' exciter's voltage available to the a-c generator's field for excitation purposes. A low reading will be obtained on the a-c voltmeter.

9. Manually decrease the resistance in the exciter's field by turning the exciter field

rheostat handle in the counterclockwise direction (fig. 9-32). This action

Figure 9-33.—AdjusHng the frequency.

294

increases the excitation voltage and the excitation current to the a-c generator's field ^\'inding which results in an increased a-c output voltage. Carefully watch the a-c voltmeter and continue to adjust the voltage manually until the rated voltage (4,160 volts) is reached.

10. Check the frequency by means of the frequency meter and raise or lower the prime mover's speed, as necessary, by means of the governor motor control switch until normal frequency (60 cycles) is obtained (fig. 9-33).

11. Place the voltage regulator in control by turning the voltage regulator switch to the ox or automatic position (fig. 9-34). Check the a-c voltmeter to see if the voltage reading has changed from the rated value of 4,160 volts. If it has, adjust the voltage regulator rheostat until normal voltage is obtained.

Fisure 9-34.—Placins the voliase resulator in control.

12. Release the interlock on the circuit breaker (on some switchboards the interlock is released by a manually controlled disconnect switch, on others it is released by turning the synchronizing switch to the on position). Close the generator circuit breaker (fig. 9-35). This energizes the bus and provides power to the load when the feeder-circuit breaker is closed. On metal-clad switchgear the circuit breaker is closed by mag-

Figure 9-35.—Brinsing power onto the bus.

netic action controlled by a breaker control switch. The breaker control switch is held in the "closed" position for a few seconds until the circuit breaker closes as shovm by the red indicating lamp, and then released. When the circuit breaker closes, the bus voltmeter will read the same as the generator voltmeter. 13. The generator being now connected to the external circuit, the a-c ammeter will indicate the current drawn by the load. Use the ammeter switch to check the current in each phase. Recheck the voltage across each phase.

SECURING A SINGLE GENERATOR

To secure a single generator which is connected alone to the bus, follow these general steps:

1. Reduce the generator load as much as practicable by opening the feeder breakers.

2. Trip the generator circuit breaker.

3. Turn the voltage regulator switch to off or manual positions.

4. Cut in all the resistance in the exciter field by turning the exciter field rheostat handle fuU clockwise.

5. Open the field switch slowly. Opening the switch slowly reduces the danger of generating a high induced voltage in the field circuit and breaking down insulation in the event that the field discharge resistors are inoperative.

6. Stop the Diesel prime mover by moving the operating lever to the gov. stop position.

PARALLEL PLANT OPERATION

When the load power demand increases to the point where it cannot be handled by the capacity of one generator, it will be necessary to add or parallel the output of another generator to the bus. In connecting a generator to a bus that is already enei^zed, two important factors must be taken into account. One, the incoming generator voltage must be approximately equal to the bus voltage; and two, the incoming generator voltage must have the same frequency and be in phase with that present on the bus. In order to check this, proceed as follows:

1. Start the prime mover and bring the incoming generator up to normal speed and voltage under the control of the voltf^e regulator. (Follow steps 1 through 11 as given in the section entitled single plant operation.)

2. Compare the bus voltmeter reading with that on the incoming generator voltmeter. If they are not the same, adjust the voltage regulator rheostat of the incoming generator until the incoming generator voltage is equal to the bus voltage. This is an important step since unequal voltages will cause circulating currents to exist between the paralleled generators.

3. Compare the frequency of the incoming generator with that of the bus and adjust to correspond by means of the incoming generator's governor motor-control switch.

4. Turn the synchronizing switch to the on position. The synchroscope will rotate in one direction or the other. If the switchboard is also equipped with synchronizing lamps, they will increase and decrease in brilliancy.

6. By use of the governor motor control switch, adjust the speed of the incoming generator until it is operating at approximately the same frequency as the bus voltage. This will be indicated by the synchroscope rotating slowly in the past direction and the synchronizing lamps slowly increasing and decreasing in brilliancy.

6. Make sure that the voltages of the bus and incoming generator are the same. Then, just before the the synchroscope pointer passes through the zero position (pointing vertically upward), close the incoming generator's breaker. When synchronizing lamps are used, the breaker should be closed just before the midpoint of the dark period of the lamps is reached.

After the incoming generator has been connected to the bus, there are two additional adjustments that must be made. One adjustment ensures that each generator is carrying its share of the kilowatt load. The other adjustment ensures that each generator is operating with the same power factor—that is each generator is producing its share of wattless current or reactive kv.-a. The division of kilowatt load between a-c generators operating in parallel depends on the relative setting of their engine governors, while the amount of reactive kv.-a they supply depends on the relative setting of their voltage regulators. Therefore, to adjust the generators for parallel operation:

7. Turn the governor motor control switches until the wattmeters of each generator have equal readings (if the generators have the same rating) or the load is divided in proportion to the generator ratings (if the generator ratings differ from each other). The load is increased on the lightly loaded generator by timiing its governor motor control switch in the direction that increases engine speed, while the load is decreased on the heavily loaded generator by turning its governor motor control switch in the direction that decreases engine speed. These adjustments should be made simultaneously in order to maintain a constant frequency.

8. Turn the voltage regulator rheostat of each generator until their power factor meters read the same (this indicates that each generator is sharing the burden of reactive kv.-a proportionately). If the switchboard is not equipped with power factor meters, the a-c ammeter of each generator can be used. Proper adjustment has been made when the a-c anmieters show equal currents if the generators have the same ratings; or currents proportional to the generator ratings if the generators have different ratings. The direction in which you turn the voltage regulator rheostat of each generator depends on the readings of the generator's power factor meter or a-c ammeter before the adjustment is made. The voltage should be decreased on the generator

carr3ring the largest lagging current

(lowest power factor) and increased on the generator carrying the least lagging current (highest power factor).

SECURING A GENERATOR AFTER PARALLEL OPERATION

There are two reasons why it may be necessary to shut down a generator that has been operating in parallel with the other generators in the station. One, the peak load period of the station has passed; or two, the generator plant shows evidence of erratic operation. In either case, before the generator power is removed from the bus you must perform one important preliminary operation—all the kilowatt load must be shifted from the generator being secured to the generator (or generators) still remaining on the line. This is accomplished by simply turning the governor motor control switch of the generator being secured to lower and the governor motor control switch (or switches) of the remaining generator (or generators) to raise. The wattmeters on the switchboard will indicate how the operation is progressing and when it is finished. When the load has been shifted away from the generator being secured, the generator's circuit breaker is tripped. Then shut down the plant in accordance with steps 3 through 6 in the section entitled "securing a single generator."

OPERATING RULES AND SAFETY PRECAUTIONS

The orders that you post in the station for the guidance of the watch standers should include a general list of operating rules and electrical safety precautions.

The important operating rules are relatively few and simple. They are:

1. Watch the switchboard instruments. They show how the system is operating, reveal overloads, improper division of kilowatt load or of reactive current between generators operating in parallel, and other abnormal operating conditions.

2. Keep the frequency and voltage at their correct values.

A variation from either will affect, to some extent at least, the operation of the base's electrical equipment. This is especially true of such equipment as teletypewriters or electric clocks. To maintain reasonably constant frequency, an electric clock and an accurate mechanical clock should be installed together at the powerhouse so that the operators can keep the clocks in time with each other.

3. Use judgment when reclosing circuit breakers AFTER THEY HAVE TRIPPED AUTOMATICALLY. Generally the cause should be investigated if the circuit breaker trips immediately after the first reclosure. However, reclosing of the breaker the second time may be warranted if immediate restoration of power is necessary and the interrupting disturbance when the breaker tripped was not excessive. In this respect, it should be kept in mind that repeated closing and tripping may result in damage to the circuit breaker and thus increase the repair or replacement work.

4. Don't start a plant unless all its switches and breakers are open and all external resistance is in the exciter field circuit.

5. Don't operate generators at continuous overload. Record the magnitude and duration of the overload in the log together with any unusual conditions or temperatures observed.

6. Don't continue to operate a machine in which there is vibration until the cause is found and corrected. Record in log.

The electrical safety precautions that should be observed by the station personnel are:

1. Treat every electrical circuit, including those as low as 35 volts, as a potential source of danger.

2. Except in cases of emergency, never allow work on an energized circuit. Where work on an energized circuit becomes necessary, extreme measures of • precaution must be used. Take every care to insulate the person performing the work from ground. This may be done by covering any adjacent grounded metal with in-

Bulating material such as dry wood, rubber mats, dry canvas, or even heavy dry paper in several thicknesses. In addition take the precautions of providing ample illumination; covering working metal tools with insulating rubber tape; stationing men at appropriate circuit breakers or switches so that the switchboard can be deenergized immediately in case of emergency; and making available a man qualified in first aid for electric shock.

SWITCHGEAR EQUIPMENTS INSPEQION AND MAINTENANCE

Your inspection and maintenance schedule for the switch-gear equipments in the power station will be determined for the most part by local operating conditions and the recommendations of the manufacturer. In general, a check and inspection should be made at least once every year. Particular attention should be paid to the following:

1. General cleanino. —Not only the component parts, but the switchboard itself, should be cleaned regularly. The front panels of dead front switchboards may be cleaned without deenei^ing the switchboard. Wiping with a dry cloth is usually all that is needed. Metal-clad switchgear should be opened at intervals and all internal parts inspected for surface cleanliness. Any accumulated dust should be removed, and all bus work and insulators thoroughly cleaned. A vacuum cleaner with an insulated nozzle is preferable for general cleaning purposes. However, you can use a dry cloth to wipe off the bus bars and insulating material. Cleaning live parts should be avoided because of danger to personnel and equipment. Therefore, you should exercise care to make sure that the switchboard is completely dead and will remain dead until the work is completed.

2. Electrical connections and mechanical fastening. —Inspection of electrical connections and mechanical fastenings should be carried out on a deenergized switchboard. However, don't limit yourself to just a visual inspection. Rather, touch and shake each electrical connection and mechanical part to make sure that connections are tight and mechanical parts are free to function. Pay especial attention to the bolted joints of bus bars since loose joints may cause overheating. Where necessary, use locking devices such as check nuts or lock washers to keep connections tight.

3. Meters and instruments. —Instruments and taeters should be examined to see that all are in proper condition, are registering properly, and have no cracked or broken glass, or damaged cases. Adjust the pointer of each instrument to read zero when the instrument is not being energized.

4. Indicatino lamps. —All indicating lamps should be examined, and weak or burnt out lamps replaced.

5. Instrument transformer. —Instrument transformers should be. inspected to make sure that they are in good condition, that the primary and secondary connections are tight, that the grounding connections are intact, and that the potential transformer fuses make firm contact in their clips.

6. Rheostat mechanisms. —See that there are no obstructions to ventilation of rheostats and resistors. Replace broken or burned out resistors. When replacements are not available, a temporary repair can be made by bridging burned out sections.

7. Switchgear position-changing mechanisms. —The lowering and elevating mechanisms and the pull-out

. devices of circuit breakers used in metal-clad switchgear should be inspected and tested for proper operation. Follow the manufacturers instructions for lubrication.

CIRCUIT BREAKER INSPEaiON AND MAINTENANCE

Circuit breakers should be carefully inspected and cleaned at least once every 12 months. Severe operating duty, however, may call for more frequent inspections. For example, you should certainly make a special inspection, particularly of the contacts, after a circuit breaker has opened a heavy short circuit. Also, where conditions demand that circuit breakers be kept in either the open or closed position for long periods of time, you should make arrangements to open and close them periodically, several times in succession. This will help to keep all parts in proper operating condition.

Caution is the keyword in all work performed on power circuit breakers. The following precautions should be impressed on the men over whom you have supervision:

1. Before working on a circuit breaker make sure that its control circuits are deenergized.

2. Draw-out circuit breakers in metal clad switchgear should be switched to the open position and removed before any work is done on them.

3. Disconnecting ^witches ahead of fixed mounted circuit breakers should be opened before any work is done on them. Where disconnecting switches are not provided, the supply bus to the circuit breaker should be de-energized. In addition, make sure that all breaker studs and current-carrying parts are adequately grounded before touching any part, and that this temporary ground is maintained throughout the inspection period.

Specific instructions on the maintenance and repair of the many varied types of power circuit breakers used in central power stations cannot, obviously, be included in this training course. They will be found, however, in the breaker instruction book that accompanies each set of switchgear. Draw-out type circuit breakers will prove the safest and easiest to inspect and maintain. Contacts are fully exposed

■e
X
S
o
3

3
s
.a
E
9)
I
to
s o
I
i
.a
o
c o 'S
g
> o

<v
3 73
K
.a
o
OS
a
u
a -r
5
J
"3 Si •0
9
u
o o
T
o w
OS
3
T3 oS
XI B «8
a
3
v c.

I
00
3
a
c c c ■3 o
3
o
c 08
X)
£
3
« ^ C C
S '-5
0 g
a I.
is
1 ?
o
u
B
c —

OS aj O u
o
01
O
I
c o o
OS
c
a
09
C
o
■♦J o
c c
8 E o
P c
e
0)
00
3 Jo
•3 s
- .3 O at
*>
3 _
o o o
:S i § a- "
s ^ 9
Q n
c 'E.
c o
00
u
c o
o
"V -
Sb »
oS ^
1
60
s
V
> O
3 I
306
0 tn
9

0 3
1 2
= M « .a
c S 1°
il
I'

Install larger wires; improve contact at
connections.
Install larger control transformer, check
rectifier, and be sure it is delivering
adequate d-c voltage.
Charge battery or repair or replace.

Replace blown fuse; repair faulty connection or broken wire; dress or replace
damaged contacts or clean dirty contacts in control switch.

-to 2 2
i, 1; T3
C — OJ
e o w
c -2
>. - » =: a;
a = ^
u.15
t = ^
"2 =
C >- -t^
'-^ = c
S £
SIS
Si ^
^ JS O
C'
IS
o
K
o
to c
•5
c
a
.a
. c
Ml flS c ^ o «
*2 >»

g
f S
5E = S I S §
- «
(a
"3
2 c
ts P
^ 8
^ I
I a
I.
Is
^ I
•a c8
—" c
o 8
■r >»
c Jl a;
« g
C eS
.S .2
'-S •-•
S - 5
£ c "
"2
.a
c « o
^ s ~
- .=
sS O
■S o
I
c
= o

307

by simply pulling out the breaker and drawing off the breaker-enclosing case. If necessary, a spare breaker can be used while the breaker is being inspected. In addition, test equipment designed especially for the job can duplicate all the motions and operations required of the breaker in actual service.

Generally, careful attention should be paid to the following:

Contacts should be inspected for cleanliness, severe pitting, burning, and alinement. Surface dirt, dust, or grease can be removed with a clean cloth moistened with stoddard solution. But, don't be misled by a slight discoloration of solid-silver or silver-alloy contacts. It is a normal film that does no harm and should not be removed. A black copper-oxide film on copper contact surfaces, however, calls for a cleaning with fine sandpaper (No. 2/0).

Severe pitting and burning of either the silver or copper contacts should not be allowed to progress to a point where it can cause damage to other parts of the breaker. If caught in time the contacts can be smoothed down with a clean fine file, or clean fine sandpaper. In the dressing process try to maintain the original shape of the contact surface and remove as httle material as possible. Where replacement is necessary, always replace contacts in sets rather than singly. Never use emery paper to clean

CONTACTS.

During your contact inspection, check for proper alinement and see that the contact surfaces bear with a firm, uniform pressure. When necessary, make a contact impression and adjust spring pressure in accordance with instructions in the breaker instruction book.

Your check list should include an inspection of the circuit breaker mechanism. Pins, bearings, latches, and all contact and mechanism springs should be checked for excessive wear or corrosion or evidence of overheating, and replaced where necessary.

Trip shafts, toggle linkages, latches, and all other mechanical parts of the breaker should operate freely and without binding. Check this by slowly opening and closing the

circuit breaker a few times manually. At the same time, make sure that the arcing contacts meet before and break after the main contacts. Any noted sluggishness, poor alinement, or other abnormal condition should be adjusted in accordance with the instruction book for the circuit breaker.

Arc chutes should be removed and examined for wear. If your inspection reveals broken or damaged linings, replace them with new linings. Any scale found on the surface of the chutes should not be removed without first consulting the instruction book for that particular breaker. Some' manufacturers 'wWi recommend removal, others will not.

Oil circuit breakers may also be encountered in your work. If so, you have the added responsibility of inspecting and maintaining the oil in the oil breaker tank. This inspection should be repeated every 6 months in comparison to the 12-month interval required for other breaker parts. When the insulating oil shows signs of moisture, carbonization, or dirt, it should be tested. If the dielectric strength of the oil tests at less than 16,500 volts, when tested in a standard gap, the oil should be filtered or replaced with new oil. The dielectric strength of the oil used for replacement should not be less than 22,000 volts.

QUIZ

I

1. What is the difference between a direct and indirect system for cooling of the water used in large prime movers on permanent power-plant installations?

2. How would you difiFerentlate between switchboards and switch-gear?

3. In the switchboard discussed, what voltage can be generated with the field switch open?

4. How many surge protection units would you find in the metal-clad switchgear installation serving five Diesel-electric plants?

. 5. What features identify the circuit breakers of metal-clad snitch-gear?

6. What is considered first when erecting a permanent Diesel-electric generator? When is the conduit installed?

7. List the equipment that should be grounded when constructing a permanent electric generating station.

8. What are the three main caf?egories of duties involved in a generator watch?

9. Give the general procedure for securing a single generator connected through a bus.

10. What information can you get from the switchboard instruments?

CHAPTER 10
WORKING WITH LEAD-SHEATHED POWER
CABLE
UNDERGROUND CIRCUITS

The high voltage developed at the central power station of an advanced base is usually distributed throughout the base by means of overhead feeder lines. Overhead lines are used extensively because they are comparatively easy to install and their maintenance does not require highly skilled personnel. However, you'll find that certain portions of the distribution system are made up of underground circuits. For example, an airfield is usually served by underground circuits to eliminate hazards to flying. Similarly, congested areas and material-handling areas are fed power through underground circuits.

Underground circuits usually mean the use of lead-sheathed cable, usually the multi-conductor type. A compact spacing of conductors is maintained under the lead sheathing by insulating each conductor with layers of spirally wrapped oil-soaked paper. Jute or paper fillers arc used to round out the cable which in turn is entirely covered by an oil-soaked paper jacket as protection against the outer lead sheathing. Some types of lead-sheathed cable use varnished cambric tape instead of paper as insulation.

THE PROBLEM

The enemy of lead-sheathed cable is moisture. A slight amount of moisture working its way underneath the sheath is enough to cause a vol tag J! breakdown between conductors. Therefore whenever you splice or terminate a lead-sheathed cable you must make every eflfort to exclude moisture during the operation. The p>ossible sources of moisture are: (1) perspiration from your hands, (2) water leakage through the roof or walls of the manhole or splicing chamber, and (3) a damp atmosphere resulting in condensation on the tools and materials used in the splicing operation. You can minimize these effects by: (1) wiping yom* hands with dry rags and circling j'our wrists with sweat bands, (2) placing a temporary awning or tent over the working area, and (3) keeping materials and tools in a dry place and wiping tools if they become damp.

THE SPLICER S TOOLS

The tools and equipment that you will use on a cable splicing job are shown in figure 10-1.

The HACKSAW comes in handy whenever a cutting job is required on the cable. The cable knife is used to cut away insulation and fillers after the lead sheath has been removed. The shave hook is a scraping tool used to clean and prepare the lead sheath surface for soldering or wiping. Measurements are made with the 6-foot folding rule. The eagle claw (also called a hawk bill or utility wrench) with its long handle provides the pressure necessary to squeeze on connectors or for other similar uses. The cloverleaf or cable spreader provides a means of keeping the conductors separated while you work on them (this cloverleaf is designed for a three-conductor cable). The tinner's hammer is used for dressing and shaping. The spreading stick can be put to use as a separator or wedge. Cutting of the conductors is accomplished with the 8-inch cutting pliers. The chipping knife (also called hacking knife) is used to ring and slit the lead sheath to facilitate its

Figure 10-1.—Splicer's tools and equipment.
removal. Althouffli the .sc rewduiver is not normally uRrd as a prying tool, it can be

adapted to tliis use on the soft lead sheath without damage. The catch cloth and fixi.shixg CLOTH arc pads of herringbone bed ticking used to cat eh and shape the solder during the wiping process. Tiie SOLDER POT and LADLK scrvc as container and dispenser, respectively, of the liquid wiping solder. Most of this equipment is the same as, or similar to, the tools and equipment usetl in splicing telephone cable.

The heating equiiJinent consists of the gas furxace for lieatiiig the sohler. the gas torch for local applicntion of lieat, and the oas tank for storage of the butane gas used by the torch and furnace. The long hose on the gas torch

.320005°—55-

-21

permits a connection between the gas tank on tiie sidewalk and the torch which \'^ou use in the spUcing chanjber or manliole. A regulator reduces the high gas pressure in the tank to a safe, usable pressure at the furnace or torch. If this type of heating equipment is not available, a blow torch can be substituted for the gas torch and a gasoline furnace for the gas furnace.

CABLE SPLICING MATERIAL

A typical layout of the material necessary for a splicing job is shown in figure 10 2.

The cable filling compound when poured around the completed splice drives out moisture and acts as an insulator.

Figure 10-2.—Cable splicing material.

The LEAD SLEEVE providcs a protective cover for the finished splice and when properly wiped forms a continuation of the moisture-proof lead sheathing, paper pasters applied to the sheath and sleeve limit the wiping areas of the splice and produce a neat job. The cotton tape serves as a binder to keep the ends of cut insulation in place during splicing operations (since the tape absorbs moisture, it is discarded after it has served its purpose). The varnished CAMHRic tape is used for two purposes, (1) as a protective ^Tapping over the conductors when sweating the connectors, and (2) as a replacement for the cable insulation that has been stripped away. The split copper connectors act as pressure-type joiners between the individual conductors of the cables being spliced, soldering flux (paste form) aids in tinning the copper connectors. The stick solder (50% tin, 50% lead) is used to seal the sleeve openings after the joint has been filled with compound, bar solder (40% tin, 60% lead) is the wiping solder used to seal the joint between the lead sleeve and lead sheath. The stearine CANDLE flux being milder than the soldering flux is used on the lead sheath. Finally, there is the sheet of emery cloth (Xo. 0) and the cheese cloth. The emery is put to use wherever jagged edges of solder must be smoothed down, while the cheese cloth is used for wiping the moisture fiom your hands and tools.

QUESTIONS AND ANSWERS

No doubt there are a few questions buzzing around in your mind. Most of them should be cleared up after you've read the following series of questions and answers.

Question: Do the materials illustrated in figure 10-2 represent those necessary for every type of splicing job?

Answer: Xo. The materials shown are those required for just one type of splice—a three-conductor, 4-kilovolt cable joint, with Xo. 4/0 conductors (paper insulated cable).

Question: \^^lat determines the exact material requirements for a lead-sheathed cable splice?

Answer: Primarily, the type and size of cable to be spliced.

For example if a four-conductor cable is to be spliced you will need four connectors.

Similariy, the length and diameter of the lead sleeve depends on the size of the cable and the lay of the conductors in the joint. Other requirements, such as size and amoimt of varnished cambric tape, amoimt of filling compound and taping oil, and amount of sweating and wiping solder are directly related to the size and type of cable to be spliced.

Question: Are the material specifications, mentioned in the preceding question, a matter of natural "know-how" that every cable splicer is expected to know?

Answer: Naturally as you gain more and more experience as a splicer you will retain in your mind the material specifications and other data of specific types of splices. However, your initial information \vill come from blueprints prepared especially for each type of splice. A typical blueprint as shown in figure 10-3, A and B contains (1) sectional drawings of the completed splice, (2) a table of material containing the type and quantity of splicing material needed, (3) a dimension chart list ing specific data on length of cable sheath to be removed, and length of insulation to be stripped, and (4) specific notes covering such information as ^ proper temperature of taping oil, numbers of layers of tape and other pertinent data.

THE FIRST STEP IN CABLE SPLICING

The cable pulling crew will pull the cable into the manhole. They will overlap the ends that rest on the wall cable supports so that.no excessive sags exist. However, training and racking the cables into their final positions and cutting them to the desired lengths is the initial job of the cable splicer. Cables should be trained to bring the ends as near as possible to their final positions. They are marked at the points to be joined and cut off straight across at the points marked.

I-PIECE

SECTION N-N
(CNLAReEO)
NOTES
1. TIN 4 INCHES OF SHEATH BEYOND "S." REMOVE SHEATH AND BELT INSULATION TO SPECIFIED DIMENSIONS. BIND EXPOSED BELT WITH TWO HALF-LAPPED UYERS OF 1^-fNCH TAPE.

2. CUT CEhfTER FILLER 5 INCHES FROM END OF BELT AND OUTER FILLERS AT BELT EDGE.

3. SOLDER CONDUCTORS WITH 50-30 SOLDER.

4. TAPE EACH CONDUCTOR IN FOLLOWING ORDER:

a. HALF-LAPPED LAYERS OF H-INCH TAPE TO SURFACE OF CONNECTOR AND CONDUCTOR INSULATION EXTENDING ONE LAYER OVER CONNECTOR.

b. ELEVEN HAIF-UPPEO LAYERS OF MNCH TAPE OVER CONNECTOR AND CONDUCTOR INSULATION FORMING SLOPE 3 INCHES LONG TO THE CROTCH.

5. WRAP CONDUCTORS TOGETHER FOR VA INCHES AT CENTER OF CONNECTOR WITH THREE HAIF-UPPED LAYERS OF MNCH TAPE.

FisuTC 10-3 A.—Thfee-conductor IS-kilovolf cable joint wirh No. 4/0 or 250 or 300 CM conductor (poper insulated cable).

I

SECTION M-M

Fisore 10-3 B.—Three-conductor 12-kilovolt cable joint with No. 4/0 or 250 or 300 CM conductor (paper insulated cable).

The sleeve should be prepared for soldering before it is placed on the cable. You should clean the outer surfaces and edges for 2 inches on each end and apply flux (stearine) to prevent oxidation. Place the sleeve over the cable and slide It up to a position where it will not interfere with subsequent steps in the spliciug of the cables.

318

REMOVING THE SHEATH

Carefully measure the amount of sheath to be removed as specified in the drawing. Use a sheathing hook or sheathing knife to mark the point at which a ring cut is to be made.

To make the ring cut, tap the sheathing knife lightly with the hammer; do not cut all the way through the sheath. After the sheath is completely ringed you are ready to make the longitudinal cut and remove the sheath, as shown in figure 10-4. To make this cut, liold the knife at an angle so

Figure 10-4.—Removing sheath.

that the point does not cut the insulation when it is struck with the hammer. As ypu cut the sheath, peel it away from the cable. When about 1 inch of cable sheath remains attached to the ring cut, bend the sheath back and forth to break it free without damaging the insulation. After the portion of sheath is removed eliminate the sharp edges of the cable sheath with the shave hook and tap do^v^l the edge of the sheath lightly against the insulation if the specification does not call for belling the sheath.

REMOVING BELT INSULATION

To remove belt insulation from multiple conductor cables use the following procedure:

1. Wrap three layers of \'7d^inch varnished cloth tape between the sheath and the belt insulation. Add a fourth layer to make the tape tie.

2. Unwrap the belt insulation a few layers at a time, tearing it off against the edge of the varnished cloth. Continue unwrapping and tearing until all of the belt insulation is removed. Do not cut irregular or rough edges of belt insulation with a knife. These steps are shown on figure 10-5, A and B.

Figure 10-5, A. Wrapping varnished clo^h tape between sheath and insulation. B. Removing belt insulation.

TESTING FOR MOISTURE

On impregnated paper or varnished cambric insulated cables, samples of the insulation should be removed from the end of the cable and tested for moisture. Your hands and all tools used to remove the insulation for the sample must be absolutely free of moisture or the test for moisture in the insulation will be invalid. '

To test the sample, immerse it in taping oil that has been heated to a temperature of 125** C. Bubbling or crackling of the oil indicates moisture and work on the splice should be suspended until the officer supervising work is contacted and he decides what procedure to follow.

REMOVING FILLERS

Some multiple conductor cables use fillers of such material as hemp on impregnated paper between the conductors. These fillers are handled differently on different joints. The outer fillers are cut off even with the belt insulation of lead sheaths depending on the type of joint to be made and the kind of cable used. The center is pulled out between the conductors and cut off at a specified point.

REMOVING INSULATION FROM INDIVIDUAL CONDUCTORS

To remove the insulation from individual conductors, proceed as follows: First, tie a clove hitch around the insulation to prevent unraveling. When the insulation is to be penciled make the tie about one-fourth inch from the point where the penciling is to start. With a knife or insulation , cutter make a cut around the insulation at the specified point. On single conductor high voltage cables make the cut with a hacksaw. Avoid nicking or damaging strands of the conductor. After the cable is marked start penciling about one-fourth inch from the string tie. Trim the insulation by making diagonal cuts towards the insulation being removed. Trim the insulation in a roimd uniform cone shape. Remove

the conductor insulation by sliding it off the end of the conductor or cutting it with a knife longitudinally along the conductor and peeling it off.

INSTALLING CONNECTORS

Before connectors are installed all excess compound and dirt from the surface and ends of the copper conductors should be removed with clean rags or cheesecloth. In some cases an approved cleaning solution such as stoddard, abco or other solvent with a flash point above 100° C. may be used. If the conductor is not round or strands tend to spread, bind them together with wire to facilitate installation of the connector. If necessary, round the ends of the conductor with gas pliers and hold the wires in place with No. 17 steel wire twisted tightly around the conductor about three-fourths inch from the end. If further shaping of the conductor is necessary, use the gas pliers until the connector can be slipped over the end easily.

Apply flux to the exposed surfaces of the strands. Ordinary solder flux in paste form is applied with a brush. It should be worked in thoroughly between the outer strands of the conductors so that they are covered. If the flux is in stick form, melt it against a hot ladle and permit it to drip on the surface of the conductor.

Before placing the connector on the conductor inspect the inside of it to be sure that it is weU tinned and free of excess solder so that it can be readily installed. Connectors should not be spread in order to place them over the conductor. Slip the connector over the strand of one conductor and woric it well past the center; insert the end of the other conductor into the coimector until it joins the end of the first conductor. Work the connector back so that it is centered over the ends of both conductors. If binding wires were used remove them as soon as the connector is installed. Clamp the connectors together with utility pliers as shown in figure 10-6.

To facilitate soldering and taping of multiple conductor cables, cable spreaders should be used. Insert an oval spreader just outside of the string tie that holds the insula-

Fi3ure 10-6.—Installins connectors.

lions. Pry the conductoi-s apart so that a triangular or star spreader may be installed near the end of the joints and remove the oval spreader.

To protect the conductor insulation from hot solder, roll strips of cheesecloth into rope form and wrap them around the msulation at the etlge of the connectors as sho^ n in figure 10-7. Wrap larger strips aro(md the rest of the conductor msulation. Before you begin to apply solder to the connector joint, equip yourself with goggles and make sure the ladles you are going to use for the solder are dry.

To solder the connectors use two solder ladles. With one ladle pour solder over the end of the connector and into the

Figure 10-7.—Protecting insulation from hot solder.

slot until the strand and connector are heated enough to make solder flow freely through the cable strands and out the ends of the connector, as shown in figure 10-8. Catch the excess solder in the other ladle. During this operation move the solder ladles back and forth over the end of the connector. Return the excess solder to the solder pot. Hold stick soldering flux against the hot conductor to melt it | and coat slots and end of conductors. Continue pouring solder and applying flux until the conductor is brightly tinned. Pour one ladle of solder over the connector untU the i solder becomes plastic and sets up on the connector and begins to fill the slot. When the solder reaches this stage, take a finishing cloth or piece of waste and remove the excess solder. Leave a slight excess of solder over the slot in the connector so that when the metal cools and contracts the slot will be completely filled.

Remove all excess solder and rough spots on the connector. Use a file to remove large deposits of solder. Smooth and taper the ends of the connector down to the conductor with a file and emery cloth. Do not cut into the conductor with the file or emery cloth. To complete the operation remove the cheesecloth protector strips; examine the insulation for possible burnings, trim off any charred portions of the insulation.

Conductors should not be built up with solder to fit con-

Figure 10-8.—Soldering connectors to conductors.

nectors that are over-size because conductivity of solder is not as high as copper. If it is necessary to build up a conductor to fit the connector it should be done with copper wires, copper inserts, or other approved copper material or devices.

Before you start to tape the splice, the tape should be prepared. If you have a striker or lower rated man assisting you, have him reroll the tape into smaller rolls, approximately 1 inch to 1% inches in diameter. Where hanks are more desirable have him wind the tape over his hand to form loops 4- or 5-inches long. Before the tape is used boil it out in taping oil. Place small rolls or hanks in the basket and immerse in the hot taping oil. The taping oil should be at least 110° C. Leave the tape in hot oil until all signs of foaming or indication of moisture have disappeared. Heat enough taping oil to 110° C. for use in flushing the joints during taping operation. This hot oil is used to remove any moisture, dirt, or foreign matter accumulated in the joints during the time the insulation was exposed. Before applying any tape, flush the entire joint thoroughly with hot oU to remove any moisture, dirt, or copper filings left on the insulation after soldering.

Immediately after flushing the joint, cover the surface of the cable insulation with taping oil. Apply one-half-inch half-lapped tape over the exposed strands between the connector and the penciled insulation. Continue tape up to the diameter of the connector and penciled insulation. Continue to tape up to the diameter of the connector plus one or two turns. Apply taping oil between each layer of tape.

Apply 1-inch tape over each conductor with butt wrapping up to the diameter specified for the particular joint being constructed, as shown in figure 10-9. Apply all tape with enough tension to form a solid wrapping. On small conductor cables where the depression between the conductor and the penciled insulation is shallow, eUminate the one-half-inch tape and use 1-inch tape throughout. On large conductor cables where a tapering effect is desired at the outer edges of the taping area, make the slope by applying each

Figure 10-9.—Taping joint.

layer slightly shorter than the preceding laj'er. As each roU of tape is used, lap about 3 inches off the end of the new roll under the trailing edge of the preceding roll. When you have completed the taping operation on each conductor, bind the final turn in place with a tape tie made by passing the end of the tape under the preceding turn and pulling tlie tape tight. Cut off excess tape.

Apply three half-lapped la3^ers of tape around each conductor for 2 inches to provide enough space between the conductors for compound. Apply three additional half-lapped layers of tape around all of the conductors to bind them together and strengthen the joints as shown in figure 10-10.

Figure 10-10.—Final step in taping joint.

SOLDERING AND WIPING SLEEVE JOINTS

After all taping od the splice is completed center the sleeve joint and mark the cable at opposite ends of the sleeve. Move the sleeve aside and scrape the cable sheath and coat it with flux (steariiu') for 1 inch inside and outside of the joint ends. If neutral is to be included in the end wipe or if a duct spice is to be made, scrape and flux 2 inches outside of the joint instead of 1 inch. Cut off" neutral conductor (if

Figure 10-11.—Dressing sleeve down to coble size. (A) Beginning to dress sleeve. (B) Sleeve dressed to final size.

included in the wipe to proper length) and fan out the strands to lay on the cable sheath. Tin the strands and bind them to the cable sheath with six turns of No. 6 copper wire. Center the sleeve over the splice and dress hi the ends to form a tight fit around the cable.by tapping the ends lightly with a hammer wliile rotating it around the cable as shown in figure 10-11. The angle at which the sleeve tapers down to the cable depends on the size of the sleeve and the cable involved.

Re-apply flux thoroughly to completely scraped area of the sleeve, particularl\'7d- the undercut area. Wipe the joint by stick and torch method, using ladles only where an open flame is not permitted. Melt a small amount of solder from the end of stick and apply it over the scraped area. Reheat the solder with a torch and work around the wiping surface with a wiping cloth until surfaces of cable sheath and joint sleeves are well tinned. Apply additional flux and more solder and work around the edge of the sleeve to form a tapered shoulder of solder as shown in figure 10-12.

AIR PRESSURE TEST ON A WIPED JOINT

Joints may be air pressure tested to insure against leaks. This test is especially advisable where the cable is likely to

Fisurc 10-12.—Wipins sleeve joint.

be submerged in water. Make a small slit about one-eighth inch long in the top side of the joint sleeve near one end. This can also serve as the opening to use when filling the cable with compound. Over this opening tape or clamp a valve stem having a flange curved to the cable. Use friction tape to make a tight seal. Maintain air pressure at approximately 25 pounds per square inch. Apply soap solution to all wipes and check for air bubbles indicating escape of air. If a leak is indicated remake or refinish wipe as required. Since some air may leak through a small cable use 15 pounds pressure for small cables.

FILLING JOINT WITH COMPOUND

Make a filling openmg on top of the sleeve near each end (the small slit used in making the air pressure test will do for one opening). Cut the other opening to form seven-eighths inch of a circle or a slit. The hole would be approx-imateh" three-fourths of an inch in diameter and tapered at a 45° angle. Make this cut with the point of a hook knife by tapping it lightly with a hammer, as shown in figure 10-13.

Figure 10-13.—Making filling openings.
.'isnflos*—55 22 3 29

Cornpouiids should be lieated in special containers with suitable covers. A teakettle makes a convenient container. Never heat the compound in the container in which it is received. The container with the comp)ound should never be placed directly on a lighted furnace or torch unless a metal bafflo is placed between the container and the furnace. The container is usually placed on the compound rack supplied with the furnace windshield.

Figure 10-14.—Filling joint with compound

When you heat the compound, stir it occasionally so that it is lieatod uniformly. Suspend a thermometer in an approved holder in the compound during the heating. The compound sliould ncvcr be heated to more than 150° C uidess specified by tlie manufacturer. Never use any compound for flusliing, or filling wllich shows the sliglitest contamination with dirt or pcrsists in bubbling, sizzling, or crackling when heated above 100° C. for more than 15 consecutive minutes. To fill the joint, lower one end of ii

slightly and if necessary, insert a small funnel in the filler hole in the lower end. Hold the funnel in place and seal it to the sleeve with friction tape. Pour joint filling com-poimd into the joint until it overflows at the other end, as shown in figure 10-14. Continue pouring until all entrapped air is eliminated. Allow the joint to cool for approximately one hour and add more compound if it is needed due to contraction. After the joint is cool, seal the opening by bending the lips of the filling opening back into position. Peen the edges over to make a tight seal. Make a slightly depressed surface 1^^-inch square by hammering over and around the lip. This depression is omitted in a duct splice joint because a smaller sleeve is used. Scrape the area and coat it with flux. Apply pasters around the opening to allow sufficient space for soldering. Fill the depression and tin the area inside the pasters using a soldering iron or wiping solder. Be sure that small pockets are not formed under the edges of the solder. These pockets would permit water to enter and start corrosive action.

JUNCTION BOXES

Junction boxes facilitate work on live lines; they are usually gray cast iron or aluminum alloy boxes in which several cables may be connected by means of insulated buses and supports. They are installed at street intersections or load centers and may contain fusible links to ensure against widespread service interruption. They also provide convenient means of sectionalizing the various cables and connecting in branch circuits without disrupting the mains. Junction boxes are built by several different manufacttirers adequate for almost any needs; they must be watertight when the cover is bolted in place. Machined surfaces and gasket shoes must be cleaned and in good condition. The clamping nuts must be screwed down evenly around the cover of the box and the entrance cover plates. Air pressm*e tests are frequently used to check on the tightness of any box after installation or whenever it has been opened.

INSTALLING JUNCION BOXES

I

Junction boxes are mounted on the wall of the vault with ' expansion bolts. The size and number of bolts required] depends on the size and type of box being installed. The box i should be located so that the minimum bending radius in training the cable is not less than eight times the over-all diameter of the cable. The radius is measured to the inside of the bend. Many different types of cable entrances are provided. Some use a stuffing box but a brass or lead sleeve wiped directly to the cable sheath is usually preferred. A typical installation of a cable in a stuffing box type of cable entrance is shown in figure 10-16.

Figure 10-15.—InitalkiHon of cable in a ttuffing box cable entrance.

The lead sheath and insulation are removed from the cable as required for the type of stuffing box and cable terminal being used. The exposed insulation of the cable is wrapped with two half-lapped layers of one-inch varnished cloth tape. The stuffing box is assembled in place on the cable as shown in figure 10-15, or as required for the type of stuffing box used.

A typical installation of cable in a wiped type cable entrance is shown in figm-e 10-16.

The lead sheath and the insulation are removed from the

cable for the particular type of terminal cable being used. The exposed insulation of the cable is wrapped with two half-lapped layers of 1-inch varnished cloth tape. The cable entrance is wiped to the cable sheath while it is in place on the junction box and the conductor is attached to the proper bus. The cable terminals may be soldered to the conductor or may be of some type of solderless lug. In any case, the lug is installed before the cable is taped and installed in the cable entrance. In some junction boxes the bottom is filled with a sealing compound up to the lower edge of the entrance. In others each cable entrance is filled with sealing compound.

SWITCH BOXES

Switch boxes are similar to junction boxes. They may be obtained for use with low or high voltage circuits and have the added advantage that the interconnection of the circuits can be changed without the necessity of removing the cover. A typical 600-volt three-way switch box is shown in figure 10-17. The cable entrances are similar to those used on junction boxes and potheads. They may be provided with L-shaped cable entrances for cables coming from the side, front, or rear.

POTHEADS

Many types of potheads are used for sealing the cable sheath and providing a terminal for making attachments to

IMPREGNATED MAPLE

OIL LEVEL INDICATOR I" PIPE PLUG-I
FULL FLOATING I CONTACTS vi
18: APPROX -
contacts — Pivot at this point
WIPING SLEEVE CABLEHEAO

NON-MAGNETIC PLATE

Figure 10-17.—Typical 600-volt three-way switch box.

equipment or to overhead lines. Before installation is made the pothead is disassembled and parts are wiped with a clean dry cJoth and protected against dirt, moisture, and exposure. A disassembled pothead is showTi in figure 10-18. The body of the pothead is thoroughly inspected inside for burrs which must be removed. The wiping sleeve is cut to fit the cable and all burrs are filed off. The outside of the wiping sleeve is tinned and the sleeve and gasket are put over the cable out of the way of the work. The pothead body is then mounted on the pole or structure at the point shown on the construction drawing. The cable is trained to the final position so the radius of any bend is not less than eight times the outside diameter of the cable. Enough cable must be available to permit installmg it in the pothead according to detailed construction drawings or manufacturers' instructions. If the cable is to be attached to the pothead with a wiping sleeve a reference mark is made on the lead sleeve opposite the point where the edge of the wiping sleeve is to be located. If the terminal is to be installed on metallic armored lead-covered cables, armor tapes and jute coverings are removed for a few inches beyond the point at which the lead sheath is to be removed. The armor tapes are wiped to the lead sheath with wiping solder. The lead sheath and insulation are removed and the insulation is tested for moisture as previously described in the procedure for splicing cable.

INSTALLING POTHEADS

Figure 10-19 illustrates the installation of a three conductor 4- and 12-kUovolt nondisconnecting pothead. The discussion of pothead installation will he limited to this example. Methods of installing different makes and types

see OCTML
FOR OVERHCAO TCfMMNM.

4"STEEL PIPE RISER
AERIAL Mie
NO.OTO 350 CONNECTOR

Fisur« 10-19.—Three conductor 4- and 12^ilovolt nonditconnectins porheadt.
336

of potheads will vary, but a thorough study of this section together with practice and reference to manufacturer's instructions and specifications should enable you to install any pothead. To install the pothead shown in figure 10-19 proceed as follows:

1. Be sure at least 3 inches of cable extend above the porcelain tube after the pothead is mounted and cable is trained into final position. Remove the sheath and belt insulation to the dimensions specified.

2. Mark each conductor at a point coinciding with the top of its respective porcelain tube. Determine dimension X shown in the detailed drawing for the overhead terminal for the specific pothead and cut conductors at this dimension below the mark.

3. Tin conductor strands and inner surface of stud. Solder the studs by feeding solder through soldering holes while the stud is heated with a blowtorch.(See fig. 10-20.)

Figure 10-20.—Soldering studs in place. 337

4. Wrap the conductors with tape in the follo\%'ing order applying taping oil to each layer. First, use half-lapped layers of)^ineh tape to tape the surface of the stud connectors. Next, use half-lapped layers of 1-inch tape over each conductor, starting at the crotch and lapping over the stud connector, apply 4 layers for 4 kilo volts and 6 layers for 12 kilo volts.

5. Bind the connectors together with six half-lapped layers of 1-inch tape.

6. Flush the conductors with taping oil heated to 100° C.

7. Mount the pothead body with the wiping cap in position. Solder the cap to the sheath and install filling pipe, as shown in figure 10-21. Assemble cap nuts loosely. Fill the pothead with compound heated to 130° C. allowing the compound to rise to the vent opening. Seal the vent and continue filling until 1 pint overflows. Allow the compound to cool for 2 hours. Tighten cap nuts and remove filling pipe and

Figure 10-21.—Soldering cap to sheath.

seal with plug painted with sealing lacquer. Fill in with rubber tape where necessary. Wrap over all with two half-lapped layers of rubber tape, then with two half-lapped layers of friction tape. Paint with weatherproofing compound.

MAINTENANCE

Junction boxes, switch boxes, and potheads should be inspected once each year. This, inspection should be made in conjunction with the inspection of the other parts of the underground distribution system. After making sure that the cases of junction boxes, switch boxes, and potheads are grounded, they should be checked for temperature by feeling with the bare hands. Since this equipment should not run warm an investigation should be made to determine the cause of heat if a piece of equipment is warm to the touch. Heating may be caused by loose or corroded contacts or connectors, or by moisture causing leakage of current. If junction boxes operate with an internal air pressure they should be checked monthly to insure that the air pressure is maintained.

The bushings of the potheads should be cleaned and inspected to make certain that they are not broken. Minor chips and hairline cracks that do not extend through the porcelain glaze may be disregarded. If large chips are not covered by the top of a disconnecting-type pothead, cracks extending through the porcelain glaze are likely to cause leakage to the case. The porcelain on the pothead should be replaced.

Checks should also be made for oil or compound leaks. They usually occur at the cable entrance. If leaks cannot be satisfactorily stopped by tightening the bolts, arrangements should be made to have the equipment removed from service for repair.

The top fittings and vent plugs of potheads should be removed and checked for water. Small di'oplets of water may be caused by normal condensation. Greater quantities should be reported to the electrical shops officer who may

replace the pothead. If the compound level is excessively low so that the conductors are exposed, additional compound should be added. Since the compound expands and contracts with temperature changes, care must be taken when replacing the compound to avoid excessive pressure when the pothead warms up.

The cable sheath from the duct entrance or riser pipe to the junction box, switch box, or pothead, should be checked for cracks, corrosion, or a pitted condition. •

QUIZ

1. What purpose does cable filling compound serve in splicing cable?

2. How may insulation be tested for moisture?

3. What safety precautions should be observed when soldering connectors?

4. How should a conductor be built up to fit a connector?

5. What temperature should the taping oil be for boiling out tape?

6. What air pressure should you use to test small cables?

7. What is the maximum temperature for heating filling compound?

8. In training a cable to a junction box, what should be the maximum bending radius of the cable?

9. What condition (s) may cause junction boxes, switch boxes, or pot-heads to heat up?

APPENDIX I ANSWERS TO QUIZZES

CHAPTER 1 SEHING YOUR SIGHTS

1. You can make a group of men feel that their work is important by showing them that their job is an essential part of something big.

2. Three characteristics of an easy-to-follow order are: (1) it is clear; (2) it is concise; (3) it is complete.

3. A supervisor should delegate authority when he is overburdened with detail or when the number of men, the time factor, or the distance, limits his control over a situation.

4. A man to whom authority is delegated should understand what is to be done and the extent of his authority.

5. Four character traits that help in getting along with people are: (1) dependability, (2) punctuality, (3) consideration, and (4) tact.

6. The factors to be considered in setting up a job plan are: (1) the job; (2) how to accomplish it; (3) how many men are required; and (4) what tools and materials are necessary.

7. When an instructor does not know the answer to a question asked by a student, the instructor should admit that he does not know and ask if anyone in the class knows the answer; if that fails he should ask a class member to look it up or do it himself.

8. The man in charge of a work detail is responsible for the safety of the men.

9. United States Navy Safety Precautions, OpNav 34P1, discusses saifety precautions for naval personnel.

CHAPTER 2 ADMINISTRATIVE DUTIES

1. The Standard Navy Stock Number should be placed on the requisition when equipment is ordered.

2. A survey board is appointed by the commanding officer.

3. The functional component system is designed to ensure that a newly established base can begin operations.

4. Design Criteria for Advanced Base Construction contains basic information pertaining to construction at advanced bases.

5. Technical explanations of electrical phenomena are found in engineering references.

6. Navy training courses are prepared by the Bureau of NaviJ Personnel to help enlisted personnel prepare for advancement in rating.

CHAPTER 3

WORKING WITH ELEaRICAL DRAWINGS

1. Checking for shorts, grounds, and opens. Testing the resistance value of the insulation, splices, insulators, etc.

2. To keep freshly poured concrete out of conduits, the ends of the conduits should be plugged.

3. No. BuDocks has standard draTiings.

4. (1) The equipment is installed and connected according to the plan and specifications.

(2) The equipment is installed according to requirements of the electrical code.

(3) Standard installation methods are followed in accordance vitb the requirements for each job.

5. The plot plan.

6. The title block contains the drawing title, its date, and its number.

7. The wiring diagram should be checked before a transformer connection is made.

8. Three standard drawings are required for a machine shop.

0. The bill of material on each reference drawing shows all items necessary to install a particular portion of the wiring.

10. Adapters are used to mount receptacles, sockets, and racks in Butler buildings.

CHAPTER 4

ELECRICAL REPAIRS

1. The layout of an electrical repair shop is determined by the stt and mission of the base.

2. The commanding officer has authority to order changes in shop layout suggested by advanced base drawings.

3. The goal of any inspection schedule is to keep the station's equipment in good condition with the minimum effort.

4. Two general rules for successful motor and generator maintenance are (1) keep it clean, (2) keep it lubricated.

5. For best op)eration of a direct current machine, the brushes should be set on the commutator at the neutral point.

6. The two abrasives that niay be used in seating brushes are (1) sandpaper, (2) brush seater.

7. Carbon tetrachloride should not be used for routine cleaning of the commutator because of its corrosive effect and toxic properties.

8. Two reasons why a light cut should be taken when a commutator is turned down in a lathe are: (1) a heavy cut may cause the commutator to be ground into a noncylindrical shape, (2) a heavy cut may cause the turning tool to twist the commutator bars and to cut deeper at one end than at the other.

9. Five troubles common to a-c and d-c motors are: (1) motor fails to start, (2) motor vibrates or makes excessive noise, (3) bearings overheat, (4) windings overheat, (5) motor burns out.

10. Two practical tests for locating shorts, opens, and grounds in armatures are the growler test and the bar-to-bar test.

11. Before an armature that may need to be rewound is stripped, all available winding data should be recorded on a card.

12. When coil terminals are being soldered to the commutator, tipping the armature prevents the solder from flowing down the back of the commutator.

13. If it is necessary to rebuild a commutator, molding micanite should be used to insulate between the spider and the commutator.

14. Moisture causes most grounds on armatures and coils.

15. The purpose of insulating varnish on an armature is to fill the fiber of the insulating fabric for waterproofness, dielectric strength, and flexibilitj'.

16. The two general classes of field coils are shunt and series.

17. Before a field coil is unwound, the following data should be recorded: (1) dimensions of the coil, both with tape on and tape removed, (2) weight of the coil without tape, (3) size of wire, (4) kind of insulation.

18. Dielectric strength for a field coil is attained by winding the coil with linen tape, dipping the coil in baking varnish and baking for 10 hours at 125° F.

19. After a coil has been wound, taped, formed, dipped, and baked, it must be tested for polarity, grounds, opens, and shorts.

20. After a stator has been rewound and connections have been made, the stator should be tested for grounds, opens, shorts, and reversed connections.

CHAPTER 5 I INSTALLING COMMUNICATIONS EQUIPMENT \

1. The two basic types of telephone systems are dial and manual. |

2. The source of energy for a local-battery telephone is a dry-cell \ battery and a hand generator. |

3. A drop signal is an electromechanical device with a shutter which faUs and attracts the attention of the operator when the user of a telephone signals.

4. The stations in a common-battery system obtain the electrical energy for transmission and signaling from a source of power at the central office.

5. Automatic signal lamps are used to signal the op>erator in a common^ battery system.

6. The line jack telephone circuit in the 12-Iine switchboard is commonly called the line pack.

7. The operator's telephone circuit in the 12-line switchboard is commonly called the operator's pack.

8. A maximum of 29 lines may be served when two 12-Une switchboards are stacked.

9. On a 12-line board, the failure of a drop to function is probably caused by a defective line pack.

10. A switchboard containing both local- and common-battery lines is designed for a base where the majority of telephones are in a centrally located area, but where a few are in an outlying area.

11. A main distributing frame serves to terminate the outside lines and to connect them to the proper lines on the switchboard.

12. The main auxiliary equipment for the 57-line board is a power service panel, power unit, junction box, accessories, rectifier, storage batteries, and control panel.

13. The switchboard controls the location of other equipment in the telephone system.

14. The conductivity of the earth surrounding the ground rods can be improved by moistening it with salt water.

15. The four basic operations of a preventive maintenance progr*™ for switchboards are

(1) inspecting, (2) tightening, (3) cleaning, and (4) adjusting.

16. The portions of the 57-line switchboard system that should be inspected daily are the switchboard exterior, night alarm belL headset, chestset, exterior of MDF, storage batteries, and the control panel.

17. The first step in effecting repair in a telephone system is to determine the probable cause of trouble.

18. The basic parts of an intercom system are one or more master stations, a junction box, one or more remote speaker units, and the wire necessary to make the connections.

19. Insp>ection is the most important operation in a preventive maintenance program.

20. Public address systems at advanced bases are used for auditoriums, outdoor movies, or camp communications.

21. The most important point to remember in searching for faults in a public address system is that the trouble should be localized and isolated.

CHAPTER 6

TELEPHONE CABLE SPLICING

1. Telephone cable wire is usually insulated with paper.

2. High dielectric insulation is the name sometimes given to double-wrapped paper insulation.

3. A short length of bonding wire or ribbon soldered to the two pieces of cable will prevent creeping of the cable caused by temperature changes.

4. Cable ends should not be left open more than 24 hours without being soldered.

5. To solder the end of a cable closed, a soldering iron and 50-50 solder or an acetylene or blow torch and a stearine core solder should be used.

6. Large cable should be boiled out after each 200 pairs are spliced.

7. Electrician's scissors should be used only for cutting small gage wires or for skinning wrapping paper and pulp insulation.

8. Bridge splicing is the process of joining three or more wires together to form a bridge.

9. Splices are dried by the use of a desiccant or by boiling out with hot paraffin.

10. When a desiccant is used, the length of time required for a splice to dry depends on the size of the splice, the tightness oU the binding, and the type of insulation.

11. Anhydrous calcium sulphate and silica gel are the principal desiccants used in drying splices. '

12. Wiping metal should be approximately 40 percent tin and 60 percent lead.
326605'—55 23 3 45

13. The six steps in wiping a joint are (1) tinning, (2) heating, (3) rough forming, (4) wiping, (5) cooling, and (6) inspection and removal of pasters.

14. When a joint being wiped is poorly heated, small amounts of chilled metal appear at the paster edges.

15. A cold joint is a rough and chalky joint which gives the i^pearance of having been wiped with cold metal.

16. The cooling process of a wiped joint can be hastened by applying stearine.

17. Silk- and cotton-insulated cables being prepared for splicing are boiled out in beeswax or a special petroleum wax.

18. In splitting cable to make minor repairs, the small blade of tbe cable stripper should penetrate from %t to M« inch.

19. Repairs made with an acetylene torch and stearine core solder should extend \'7diio %

inch beyond the defect.

20. The Army manual Lecul ShecUh Telephone Cable Splicing, TM 11-372, contains information on cable splicing.

CHAPTER 7

TRANSFORMER INSTALLATION AND REPAIR |

1. Make a rough drawing of the area. ;

2. In the electrical center of the area.

3. On the ground or on a platform.

4. 100 amperes. j

5. They are merely placed on the platform. Their weight keeps them | in place.

6. In addition to the normal grounding of neutral wires, and lightning arresters, all equipment and steel must be effectively and permanently grounded. Protective fences must be built around the installation and warning signs prominently posted.

7. Broken bushings.

8. (1) Removing the internal wiring leads from the defective bushings.

(2) Replacing the clamping device that holds the defective bushing in place.

(3) Removing the defective bushing.

(4) Inserting a new bushing.

(6) Clamping the new bushing in place.

(6) Connecting the internal leads to the new bushing.

9. It provides insulation between the windings and prevents high voltage arc-overs. In addition, it aids in cooling the transformer by conducting the heat from the coil to the transformer case.

CHAPTER 8

ADVANCED BASE PORTABLE ELEQRIC POWER

PLANTS

1. An internal combustion prime mover and alternator and a switchboard.

2. 9 part« water, 20 parts acid.

3. A solution of bicarbonate of soda.

4. (1) Check all electrical connections with the plant's connection diagram.

(2) See that the connections are tight.

(3) See that the collector rings are clean and have a polished surface.

(4) Check collector brushes to make sure they have no tendency to stick in the brush holders, that the^ are properly located, and that the pigtails will not interfere with the brush rigging.

(5) Check the collector brush pressure to see if it agrees with the figure recommended in the plant's instruction book.

(6) Check the exciter in the same manner as the alternator.

5. Load current to be carried, and distance beween the load and the plant.

6. A pipe over IH inches in diameter.

CHAPTER 9 CENTRAL ELECTRIC POWER STATIONS

1. In the direct system the heat of cooling water is pumped to a spray or cooling tower where it is cooled by exposure to open air and by evaporation. In the indirect system the cooling water is recooled by means of a radiator or heat exchanger.

2. Switchboards are the panels or control desks on which meters, relay instrument,

switches and control handles are mounted. Switch gear is a coordinated assembly of switchboards and switching equipment necessary for the control and distribution of the electrical power produced in the central power station.

3. Only that voltage due to the residual magnetism of the generator field structure.

4. One.

5. The circuit breakers of metal-clad switch gear are removable and interchangeable as a unit.

6. The foundation; before the concrete is poured.

7. (1) The frames or metal base of each electric motor and generator.

(2) The metal cabinet of the power panel board.

(3) The neutral bus in the metal-clad switch gear.

(4) The fuel tanks and transformers.

(5) The field posts that support the chain hoist over each plant.

(6) The building frame if it is of metal.

8. (1) Operating the station's equipment.

(2) Maintaining the station's equipment.

(3) Keeping a daily operating log.

9. (1) Reduce the generator load as much as practicable by opening the feeder breakers.

(2) Tip the generator circuit breakers.

(3) Turn the voltage regulator switch to off or manual position.

(4) Cut in all the resistance in the exciter field by turning the exciter field rheostat handle full clockwise.

(5) Ojjen the field switch slowly. Ojiening the switch slowly reduces the danger of generating a high induced voltage in the field current and breaking down the insulation.

(6) Stop the Diesel prime mover by moving the operating lever to the governor stop position.

10. How the system is operating, overloading, improper division of kilowatt load, of reactive current between generators operating In parallel, and other abnormal conditions.

CHAPTER 10

WORKING WITH LEAD-SHEATHED POWER CABLE

1. The cable filling compound drives out the moisture and acts as insulation.

2. Immerse it in taping oil that has been heated to 125** C. Bubbling or cracking of the oil indicates moisture in the insulation.

3. Wear goggles and be sure that the ladles you use are dry.

4. The conductor should be built up with copper wire, copper insert, or other approved copper material.

6. 110° C.

6. 15 pounds per square inch.

7. 150° C.

8. Eight times the over-all diameter of the cable.

9. Heating may be caused by loose or corroded contacts or connectors, or by moisture causing leakage of current.

APPENDIX It

OPENINGS AND SLEEVE DIMENSIONS FOR STRAIGHT TELEPHONE CABLE SPLICES

22 Gauge Aerial Straight Splices

19 Gauge Aerial Straight Splices

Main Branch

350

Cut)

351

Main Branch

352

Not Cut)

353

I

APPENDIX III

STANDARD CABLE COLOR CODE

Singles

Pairs

Fairs —Continued

Pairs —Continued

Pairs —Continued

Triples

Spare Singles

Num- Number Color her Color

1 Red-White 3 Red-Black

2 Black-White 4 Red-Black-White

Spare Pairs

APPENDIX IV

QUALIFICATIONS FOR ADVANCEMENT
IN RATING

CONSTRUCTION ELEORICIAN'S MATES (CE)

Ratins Codt No. 5300

Scope

General Service Ratios

Construction electrician's mates install, operate, maintain, and repair electrical generating equipment, distribution systems (primarr and secondary), transformers, switchboards, distribution panels, motors, controllers, and master switches, interior wiring, and lighting fixtures; erect poles, attach insulators, string wires, and lay cable for electrical distribution and telephone lines; install, operate, maintain, and repair PBX exchanges, switchboards, telephones, public address systems, and interoffice communication systems.

Emergency Service Ratings

Construction Electrician's Mates G (General Elec- ;
tricians), Rating Code No. 5301 CEG

Work on ground and in shop on all types of electrical I installations.

Construction Electrician's Mates P (Power Linemen),
.Rating Code No. 5302 _.. CEP

Work aloft and aground on electrical distribution systems.

Construction Electrician's Mates L (Communications
Linemen), Rating Code No. 5303 CEt |

Work aloft and aground on telephone, public address, and interoffice communication systems.

Navy Enlisted Classificotion Codes I

For specific Navy enlisted classifications included within this rating-see Manual of Navy Enlisted Classifications, NavPers 15105 (Revised), codes CE-5600 to CE-5699.

Qu<iiifications (or Advancement in Rating

Qualifications for Advancement in Rating

100 PRACTICAL FACTORS

101 Operational

1. Select and use common electrical and communication hand tools

2. Operate and parallel advanced-base d. c. generating equipment; stand generator watches

3. Solder electrical connections and wire splices

4. Read wiring diagrams and sketches

5. Perform the following under simulated conditions:

a. Rescue of person in contact with an energized circuit

b. Resuscitation of a person unconscious from electrical shock

c. Treatment for electrical shock and burns..

d. Pole-top resuscitation _

6. Install, under supervision, interior wiring and electrical fixtures. Bend and install conduit

7. Operate and parallel alternators, using either synchroscope or synchronizing lamps; stand generator switchboard watches

8. Install storage and dry-cell batteries used for communication systems

9. Operate manual telephone switchboards at advanced bases

10. Install, under supervision, interior telephone wiring, subsets, and bell and buzzer circuits

11. Work in a crew erecting pole lines and accessories for:

a. Electrical distribution lines

b. Telephone lines

12. Read and work from electrical plans and wiring diagrams

326605"—55 24 361

AppiicabU Pale*

CE CES CEP CEl

3
3
3 3
3
3
3 3
3 3 3
3 3
3 3
3 3
3 3
3

3

3 3

Qualifications for Advanoement In Rating

101 Operational—Continued.

13. Install power transformers:

a. On the ground

b. On poles

14. Install interior telephone wiring, subsets, and bell and buzzer circuits

15. String telephone lines. Pull in overhead and underground telephone cables

16. String electrical distribution lines

17. Install advanced-base-type generators, switchboards, and distribution panels

18. Install current and voltage transformers.

19. Lay and splice cable for electrical distribution systems __

20. Lay out and install secondary electrical systems.-

21. Operate lathes to turn down commutators

22. Install the following at advanced bases:

a. Manual telephone switchboards

b. Public address systems

c. Interoffice communication systems

23. Splice overhead and underground telephone cable. Splice-in cable terminals. .

24. Stand watch in control room of an electric generating station

102 Maintenance and/or Repair

1. Use voltmeter, ammeter, and ohmmeter in checking and testing circuits and equipment. Locate and repair grounds, open circuits, and short circuits, under supervision

2. Charge storage batteries. _

3. Clean and lubricate electric motors and advanced-base generating equipment

4. Maintain and repair interior wiring and lighting fixtures

ApplieabU Ptttt

CE CE8 CEP

2

2

1 1

1

1

1

1 1

1

1

C

Qualifications for Advancement in Rating

6. 7.

8.

102 Maintexaxce and/or Repair —Con.

6. Maintain electric motors and advanced-base generating equipment, excluding prime movers

Rebuild storage batteries

Repair grounds, open circuits, and short
circuits in electrical systems
Maintain and repair interior telephone wiring and bell and buzzer circuits.
Check and maintain subsets
Maintain and repair current and voltage
transformers
Reimir advanced-base electric generating and control equipment, excluding
prime mover
Analyze and correct faults in electrical
equipment and circuits
.\nalyze and correct faults in telephone, public address, and interoffice communication
equipment
Maintain and repair the following:
a. Electric distribution lines
b. Electric motors and generators
c. Electric relays, .solenoids, switches, circuit breakers, fuses, and controllers..
Rewind, insulate, and bake armatures
and fiokl coils
15. Maintain and repair the following:
a. Telephone pole lines and wiring systems
b. Manual telephone switchboards
c. Public address systems
d. Interoffice svstems
10
11
12
13
Applicable Fate$
14
103 Administrative and/or Cleuicai,
1. Make material estimates from electrical working drawings, sketches, and specifications
2. Prepare re\'7b|uisitions for electrical materials or e(|uipment
CE CEG CEP GEL
3 2
2 1
1 1
1 1
2 2
Qaallflcations (or Advancement In Rating
103 Administbative and/or CLEBicAii—Con.
3. Supervise and train personnel engaged in the installation, maintenance, and repair of:
a. Telephone, public address, and intercommunication systems
b. Electric distribution system
0. Electric wiring and equipment
4. Organize and manage an electric shop
and an electric generating plant

ApplieabU Fata

200 EXAMINATION SUBJECTS

201 Operational

1. Principles of magnetism as applied to a. c. and d. c. motors and generators

2. Construction features of a. c. and d. c. motors and generators

3. Types, characteristics, and uses of a. c. and d. c. motors and generators

4. Electrical symbols used on working drawings and wiring diagrams

6. Common properties of electric circuits (resistance, inductance, capacitance) and units of electricity (volts, amperes); Ohm's law

6. Methods of rescuing a person in contact with an energized circuit. Resuscitation of a person unconscious from an electrical shock

7. Treatment of electrical and acid bums___

8. Safety precautions to be observed when working with or in vicinity of electric circuits and equipment

9. Operation of manual telephone switchboards _

10. Safety precautions to be observed when:

a. Handling molten lead compound and taping oil used in cable splicing

b. Climbing poles, buildings, or towers to string or work with wires and cables.

o. Mixing electrolyte and charging batteries

CC , CE6

C

c

C

S

C C

s

3

3

3

3

3

3 3

CEP ca

3

Qoaliflcations for Adranoement in Rating

201 Operational —Continued.

11. Safety precautions to be observed in installation, operation, and maintenance of electric circuits and equipment for:

a. General wiring and power systems 3

b. Communication systems 3

12. Method of administering pole-top resuscitation 8

13. Tools, materials, and equipment used in the installation, repair, and maintenance of:

a. Electric distribution systems 3

b. Electric motors and generators 3

c. Telephone pole lines and wiring systems _

d. Public address and interoffice communication systems \ 3

14. Uses and types of electric circuit break- I ers, controllers, switches, relays, sole- | noids, and fuses 2

15. Advantages, characteristics, and uses of '
Y. delta, and v connections | 2

16. Types of manual and automatic telephone equipment used at advanced bases

17. Relationship between generator capacity, | load, and distribution system capacity ;

18. Types of public address and interoffice ; communication systems

ApplkaNe Raie$

CE

202 Maintenance and/or Repair

1. Methods of maintaining, rebuilding, and charging storage batteries

2. Methods of maintaining and repairing the following:

a. Electric distribution lines

b. Electric motors and generators

c. Electric relays, solenoids, switches, circuit breakers, fuses, and controllers.

CE8 CEP

2 2

Qaaliflcations for Advancement In Rating

202 Maintenance and/ob Repair —Con.

3. Methods of maintaining and repairing the following:

a. Telephone pole lines and wiring systems. _

b. Manual telephone switchboards

c. Public address systems

d. InteroflSce systems

203 Adminibtbativb and/ob Clerical

1. Functions of operational companies of a Construction Battalion

Applicable Rda

CE CEa CD

1 1 1 1

ea

U. S. eOVERHMENT PRINTINS OFFfCE: I9SS